ENGINEERING MECHANICS

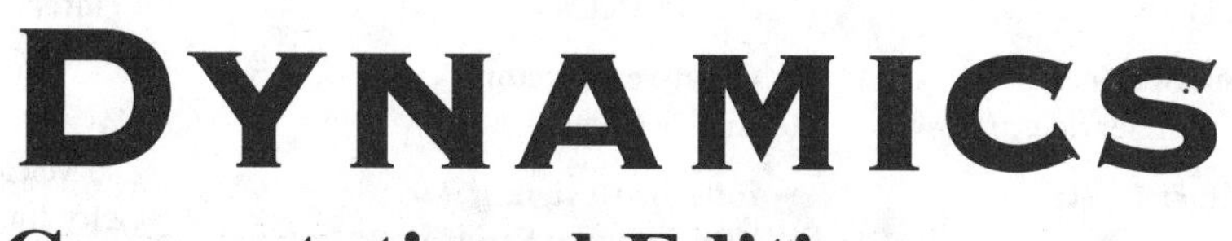

Computational Edition

Robert W. Soutas-Little
Michigan State University

Daniel J. Inman
Virginia Polytechnic Institute and State University

Daniel S. Balint
Imperial College London

Australia · Canada · Mexico · Singapore · Spain · United Kingdom · United States

THOMSON

A MATLAB® Manual for Engineering Mechanics: Dynamics, **Computational Edition**
by Robert W. Soutas-Little, Daniel J. Inman, and Daniel S. Balint

Publisher:
Chris Carson

Developmental Editor:
Hilda Gowans

Permissions Coordinator:
Vicki Gould

Production Services:
RPK Editorial Services, Inc.

Copy Editor:
Pat Daly

Proofreader:
Erin Wagner

Indexer:
Shelly Gerger-Knechtl

Production Manager:
Renate McCloy

Creative Director:
Angela Cluer

Interior Design:
RPK Editorial Services

Cover Design:
Andrew Adams

Compositor:
International Typesetting and Composition

Printer:
Thomson/West

Cover Image Credit:
Zac Macauley/The Image Bank/Getty Images

Printed and bound in the United States
1 2 3 4 07

For more information contact Thomson Learning, 1120 Birchmount Road, Toronto, Ontario, Canada, MIK 5G4. Or you can visit our Internet site at http://www.thomsonlearning.com

Library Congress Control Number: 2007927300

ISBN-10: 0-495-29608-2

ISBN-13: 978-0-495-29608-9

North America
Thomson Learning
1120 Birchmount Road
Toronto, Ontario MIK 5G4
Canada

Asia
Thomson Learning
5 Shenton Way #01-01
UIC Building
Singapore 068808

Australia/New Zealand
Thomson Learning
102 Dodds Street
Southbank, Victoria
Australia 3006

Europe/Middle East/Africa
Thomson Learning
High Holborn House
50/51 Bedford Row
London WCIR 4LR
United Kingdom

Latin America
Thomson Learning
Seneca, 53
Colonia Polanco
11560 Mexico D.F.
Mexico

Spain
Paraninfo
Calle/Magallanes, 25
28015 Madrid, Spain

Contents

Introduction

This supplement to *Engineering Mechanics: Dynamics* Computational Edition by Soutas-Little, Inman and Balint, will provide all the necessary instructions to use recent versions of MATLAB software, and in particular the *Student Version* of MATLAB available from the web at http://www.mathworks.com/academia/student_version/, to aid in solving the homework problems and working through the sample problems. It should be noted that before any attempt is made to use this computational software, the problem must be correctly modeled, that is, a free-body diagram constructed and the correct equations of motion formulated. Computational software only reduces the numerical burden, makes it possible to visualize the solution through plots, and facilitates parametric studies which enable design and physical principles to be examined. MATLAB is a registered trademark of The Mathworks, Inc., 3 Apple Hill Drive, Natick, Massachusetts, 01760. Both the *Student Version* and Professional versions are available from the Mathworks. Many institutions have a site license for student access and the *Student Version* is often available in campus bookstores. The web address listed above allows a free download of the manual for MATLAB, if more detailed instructions are of interest. The instructions for the student and professional versions are the same. Just the prompt is different as described below.

A large number of universities have fantastic web sites introducing the use of MATLAB. If you run in to trouble understanding any aspect of the code the answer can often be found by simply typing "matlab" into Google which will reveal a host of excellent sources. In addition, the Help file in MATLAB is also very useful. The Mathworks web site is the best source for using the code as the Mathworks continually updates their software and their web site.

This supplement is intended to guide the reader through the use of MATLAB for solving dynamics problems. As such, you are encouraged to open your version of MATLAB and try the various steps described in the following as you read through it. It is keyed heavily to the accompanying text and works through many of the sample problems in detail, solving the sample problems from the text using MATLAB for each chapter. It is suggested that you work through the supplement with MATLAB open on your computer until become comfortable using the program. Then refer to the supplement as needed in solving the homework and studying the text. This supplement is written in a style consistent with the *Student Version* as used in most MATLAB documentation and is keyed to the *Dynamics* text material. While this supplement suggest ways to use MATLAB to enhance your understanding of Dynamics and teach you efficient computational skills, you may browse through the *Student Version* or professional version manuals and think of your own usage of MATLAB to solve dynamics problems. A key benefit of using MATLAB is the ability to visualize the motion that results from the solutions to dynamics problems as they evolve in time.

Wherever possible, the input to, and the output from Matlab has been copied and pasted directly into this text. The fonts however, have been modified to make clear exactly what the reader should type in order to reproduce the results discussed in this manual. There are three fonts used here. The font for the base text which provides information and discussion, the font to indicate that it is MATLAB generated: `EDU>`, and the font indicates information that is entered into MATLAB by the user: x=[35; 45; 10]. In some cases the spacing has been changed to "fit" the page and to avoid large empty spaces. Note the prompt for the student version is `EDU>`, while the prompt for the professional version is >>.

This supplement consists of nine sections. The first section is an introduction to using MATLAB, concluding with an explanation of its use as a simple calculator. If you have worked through the *Statics* version or are already familiar with MATLAB then you may skip these introductory sections. The remaining chapters are keyed to those in the text (*Engineering Mechanics: Dynamics, Computational Edition*, by Soutas-Little, Inman and Balint). Each of these remaining sections presents MATLAB solutions for the Sample Problems given in the *Dynamics* text.

Introduction to the Use of MATLAB

Supplement to *Engineering Mechanics*: If you are already familiar with MATLAB or have used the MATLAB *Statics*, you may wish to skip this section.

Numerical Calculator

When using MATLAB for the first time, it is useful to think of the package as a calculator. In fact, it is worthwhile to load MATLAB and start using it to make some very simple calculations (e.g., adding, multiplying, etc.) in order to become comfortable with its format, much as you would first experiment a newly purchased calculator.

In MATLAB, numbers are entered and assigned values by using an equals sign. When you open MATLAB, the following prompt will appear inside a window called the *command window*:

```
EDU>
```

To assign the symbol "a" the value 0.24, just type `a = 0.24` after the prompt. Your screen will then look like this:

```
EDU> a=0.24
```

In MATLAB, there is no need to worry about entering spaces, as they are usually ignored (except in the case of entering data into a matrix). As soon as you hit the "return" key, MATLAB returns the result of your entry. For the example we have been using this far, you will see the following on your screen:

```
EDU>a=0.24
a =
        0.24
```

If you do not want MATLAB to write out the result of your input onto the screen, you may suppress this feature by adding a semicolon at the end of the statement (e.g. type: `a=0.24;`).

If you type any legitimate list of operations at the prompt, it will be computed when you press the "return" key. For example:

```
EDU>c=3*a^2
c =
        0.1728
```

If MATLAB cannot recognize a symbol or a command, it will give you an error message with some hint about what might be wrong.

MATLAB deals with numbers in lists, called vectors, and in arrays, called matrices. Even a scalar—or a real, nonvector number—is treated as a one-by-one matrix. Thus, if the symbol `u` is to denote a row vector of, say, the coordinates [1, -1, 2] of a particle, then to define `u`, you would type the following:

```
EDU> u=[ 1 -1 2]
u=
        1 -1 2
```

Here, columns are separated by a space (a notable exception to the general rule that spaces are ignored). If a column vector is desired, type the following, separating the entries by semicolons:

```
EDU> u=[1; -1; 2]
u=
        1
        -1
        2
```

The semicolon is used to denote a new row. Alternatively, you could also define `v=u'` to transpose the data from the row vector `u` into a column vector. The apostrophe denotes the transpose of the vector or matrix that precedes it. These commands and associated manipulations are extremely useful in mechanics, as are all of the vector manipulations (e.g., dot product, cross product, addition, and subtraction) that are also defined in MATLAB. Note that when the solution of a differential equation is computed, it is stored as a vector list of values.

The MATLAB code is based on the use of the arrays of numbers. So, it also uses this notion of a vector to keep a record of variables that have a range of discrete values. For instance, in the next example, the constant T is defined to be 10 (seconds), and the variable t is defined to start at 0, increment to 0.1, and to continue incrementing until its final value is 10, or T. Rather then entering in a list all of the values that t takes on, you can generate the list automatically by typing the following:

```
EDU>t=(0:0.1:10);
```

Here, the first number is the starting value of t, the second number is the amount by which t is to be incremented, and the last number is the ending value. These entries are seperated by colons. For instance, if the variable t was to begin at 2, have an increment of 0.1, and a final value of 8, enter: `t= (2: 0.1:8)`. Note that the semicolon at the end of the line suppresses the display of the value t. If you prefer to generate the list by specifying the number of values in the list rather then specifying the increment, type the following:

```
EDU>t=linspace(0,10,101);
```

This command produces the same list of values of `t` as does the earlier line of code. It is important to note that `t` is a row vector, each element of which is a discrete value of the variable `t` in order. Hence, `t(3)` is the third element in the vector and has the value 0.2000, and `t(101)` is the last element in the row vector and has the value 10.000.

Now that you know how to enter constants and variables and how to generate ranges of values of variables, the next step is to use these values as the domain of a function. MATLAB uses most of the standard math functions (see the inside back cover of this supplement for a list) and operates on lists contained within these functions in a single command. For example, to evaluate sin(t), just type:

```
EDU> y=sin(t);
```

This command assigns the row vector y the various values of sin(t) for each discrete value of t contained in the list for t. From this command, MATLAB assumes that you want to calculate the sine of each element in the vector t and place those values in an associated row vector y. However, a little care must be taken when multiplying and adding variables that are represented as vectors. If you need to multiply a vector variable `t` consisting of 101 elements by another vector variable `y`, `y` must contain the same number of elements as `t` does. To multiply each element of `t` by each element of a vector y that is the same size as t, type t.*y. The period before the multiplication operator signifies an element-by-element multiplication. For example, if vector t = [1 0] and the vector y = [-1 2], then typing t.*y returns ans = -1 0, which is a row vector. This answer is not a vector product of the type used in mechanics, but its calculation does demonstrate the procedure for the manipulation of variables in digital form using MATLAB.

Significant Figures

MATLAB computes all numerical answers to a high degree of numerical accuracy. You can choose how to display these answers in terms of the number of digits (5 or 16), as demonstrated on page 27 of the manual for the student edition of MATLAB, but this display is not a reflection of the number of significant digits in your answer. Thus, it is very important to interpret the results of any computer output using both the rules for significant figures and common sense. For example, if you use MATLAB to compute the average number of people driving by your home on a given day, and it returns the number 50.533 or the number 50.53333333333334, common sense tells you that this answer signifies either 50 or 51 people. The point is to remember is that the numbers returned from your commands must be further analyzed to determine their meaning.

Symbolic Math

MATLAB also allows you to make certain evaluations symbolically. For instance, you can enter a function and differentiate it symbolically. In order to let MATLAB know that a variable is to be treated symbolically rather than numerically (as in all of the previous examples, place the variable between apostrophes). For example, x denotes a numerically valued quantity, while 'x' denotes a symbolic variable. Also, symbolic objects can be created from numerical ones by using the "sym" function. For example, typing x=sym('x') creates a symbolic variable x. Once a variable is defined as being symbolic, it can be used in mathematical expressions in the same way that numerical variables can, except that the corresponding evaluation or computation is made symbolically. For example, the following set of commands differentiates the sine function:

```
EDU>x=sym('x'); % this line creates a symbolic variable x
EDU>diff(sin(x)) % this line differentiates sin(x) with respect to x
ans=
      cos(x)
```

Note that the percent sign (%) denotes that the text that follows is a comment that you type in (to help you or someone reading your solution to understand your code) and is not part of the code. Also, MATLAB uses the notation "ans =" to denote that the answer to a command follows, unless you have assigned the calculation another name. More details on using the symbolic aspects of MATLAB appear in the rest of this text as needed.

File Saving

It is often desirable to be able to repeat a particular calculation with different values as its argument. This capability allows you to perform "what if" studies and aids in design. In the previous examples, we were working in the command window, which computes every time we hit the "return" key and has no ability to go back in our list of formulas to fix and reexecute the commands if we find that we do not like our answer or if we cause an error to occur. It often happens in engineering that after computing a long list of formulas, we look at out answer and find that it is unreasonable. For instance, if we use a string of formulas and calculations to compute the weight of a car and our answer turns out to be five pounds, we know that we made an error somewhere in our calculations. It would be useful to be able to go back and find the line that needs to be corrected, fix it, and run the formulas again without having to retype or copy and paste the entire group of commands. You can do this type of error correction by saving the file with all of your commands as a *m-file*, also called a *script file*. These special files are essentially scripts that MATLAB reads from to perform the calculations you want. The names of these files end in ".m", and hence, they are called "m-files".

To create a list of formulas as an m-file choose "New" from the File menu and select "M-file". This command will display a text-editor window where you can enter your commands. For the example we will now use, enter the following text into the window:

```
% this m-file converts inches to meters, as in Sample Problem 1.2
D=15*2.540*0.001
```

Now save the file, naming it "samp1pt2.m". Be careful not to use a dash or period in the name of the file, except, of course, to end the name with .m. Now, if you type the name of the file after the prompt (without the ".m" part), MATLAB will return the answer:

```
EDU>samp1pt2
D =
     0.0381
```

The difference between typing in the command directly and using the m-file approach is that you can go back and change the m-file if need be without retyping the entire line. For example, suppose that the diameter of a tire is measured to be 14 in., and you need to convert this measurement into meters. Then just open "samp1pt2.m", make the change, and run it again to get the new value (D = 0.03556). Effectively, the m-file allows you to edit your commands more or less as you would edit a spreadsheet so that you can make changes in your problem formulation without reentering every command. This capability is not so important in the simple example used here, but it will become important as your knowledge of mechanics increases and the level of calculation becomes more involved. This feature is the one that allows you to perform "what if" studies and to interrogate your answers to complex engineering problems. Often, long entries of data are subject to typing errors, and any dynamics problem is subject to intellectual errors (OK, mistakes). Such errors can often be found by changing a few numbers to see what happens and if the new solution makes sense. For example, in some cases, setting some parameters to zero reduces the problem to a simple one for which the analytical solution is known to you. Then, a comparison of the known analytical solution to the calculated numerical solution either gives you confidence in your computer results or sends you looking for an error in your code or in your mechanics solution. This aspect is extremely important in engineering and should become a routine part of your problem-solving technique, since the use of computational software makes recalculation particularly easy.

Defining a Function

While MATLAB has a number of useful mathematical functions defined in its library (see the inside back cover of this supplement for a list), it is often very useful to be able to define your own functions. In fact, in order to use MATLAB to solve the differential equations of motion arising in the study of dynamics, you must be able to define functions. This task can be done by creating a special m-file and using the existing functions in MATLAB.

Consider the case of a load on a beam where the load is defined by the function $w(x) = 10\ (x - x\sin(x/10))$ in units of force per unit length (say N/m). To define this function in MATLAB, use the already-defined functions and then create an m-file named after the function. For example, if we call this function "wload", then we would create an m-file (by selecting "New" and "M-File" from the File menu) called by this same name, with the first line of the m-file indicating that the function "wload" is a function of x, as follows:

```
function y = wload(x)
        % this file defines the function "wload" as a function of x
y = 10*(x-x.*sin(1/10*x));
```

We would make sure to give this file the name wload.m, for it is important that the name of the function in the file and the name of the file are the same; in fact, when you go to save the file, MATLAB will try to name it correctly for you. Note that the first line of the script must start with "function" and that in building the formula for y, we use the multiplication symbol ".*" instead of just "* " because x and sin(x) are both vectors of numbers. The "* " sign by itself is for muliplying two scalars (one-by-one matrices) together and for multiplying a scalar by a vector. If you need to multiply or divide an element by an element, use the dot notation (".* " and "./", respectively). With this special function defined and saved, you can now call up the value wload(3.5), for exmple, and get the value of the function when $x = 3.5$. If you type wload(3.5) without defining it as a function, however, MATLAB thinks that you are looking for an entry in the vector wload in the 3.5 location, which does not exist (vectors have only integer locations). The following example illustrates the use of the user-defined function wload:

```
EDU>wload(3.5)
ans =
 22.9986
```

1

Kinematics of a Particle

Inverse-Dynamics Problem

If the rectilinear displacement of a particle is known, its velocity and acceleration may be obtained by either symbolic or numerical differentiation and then may be graphed using MATLAB. This section describes how to use both of these differentiation methods and how to plot the results.

Suppose that the displacement of a particle is given to be $x(t) = e^t \sin(\pi t)$ for values of time t between 0 and 2 seconds. We will differentiate this function both numerially (which is generally a bad method to use) and symbolically. First, let us consider numerical differentiation. This type of differentiation is accomplished in MATLAB by using the function `diff` twice to compute the derivative of a function f, which is defined as an array of numbers, by performing the calculation $\Delta f/\Delta t$. The `diff` function in MATLAB takes the difference between adjacent values in an array of numbers and performs the derivative on them as applied to a symbolic function. To run the numerical differentiation, first define the array of numbers, specifying the start and stop times and the increment; next, define the function to be numerically differentiated; then use the "diff" function to compute the derivative numerically. This process is illustrated as follows:

```
EDU>t=(0:0.1:2);     % define t to be the array of numbers between 0 and 2 in increments of 0.1
EDU>f=exp(t).*sin(pi*t);          % define the displacement function
EDU>df=diff(f)./diff(t);          % compute the derivative numerically by using the ratio
                                  % of differences
EDU>dt=t(1:length(t)-1);          % reduce the length of the array t so that it is the same length as
                                  % the array df
EDU>d2f=diff(df)./dt(1:length(dt)-1);   % compute the second derivative again, reducing the
                                        % length of dt to agree with the length of df
Warning: Divide by zero.
EDU>pf=f(1:length(t)-2);pdt=dt(1:length(t)-2);pdf=df(1:length(t)-2);
    % make all of the arrays the same length for plotting purposes

EDU>plot(pdt,pf,'+',pdt,pdf,'*',pdt,d2f,'x'),title('postition,+,velocity,*,and acceleration,x')
    % plot the various differences
```

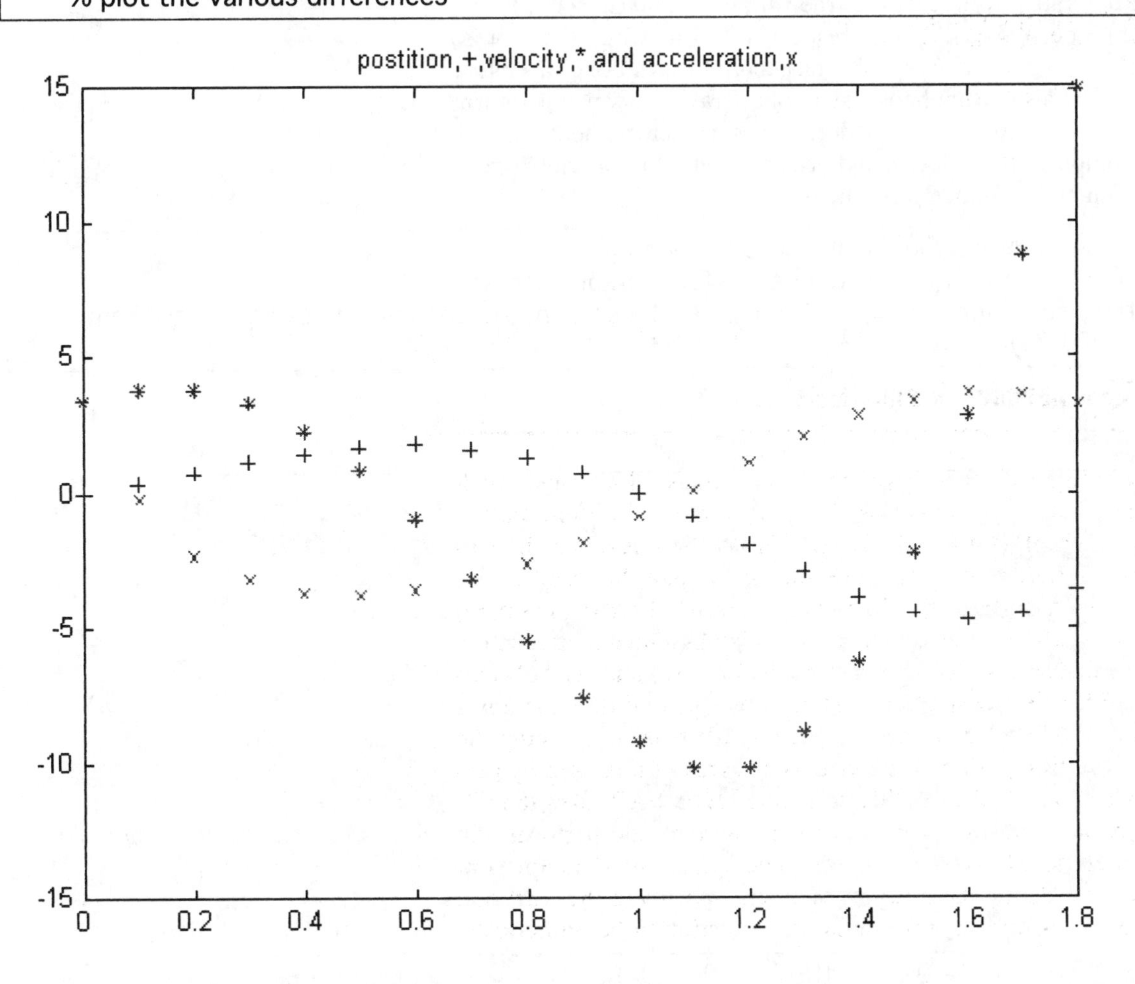

Note that in MATLAB, you must constantly be aware of the size—in this case, the length—of the numerical arrays created by your program. The length command is used to return the number of elements in a vector. In the previous example, it is used to throw away the "extra" elements in longer vectors so that all four variables are of the same length. This procedure is done because the "plot" command can only be used to plot vectors of the same length. That is, you cannot plot one array (say, the array for time) against another array (say, the array for displacement) unless the two arrays are the same size. Since the diff function produces an array that is one element shorter than the array on which it operates, the function being differentiated must be reduced by one element in order to fit on the same plot. Hence, we included the following line of code: "EDU>dt=t(1:length(t)-1);".

Next, let us consider computing derivatives symbolically. Again, the basic command we will use is diff. However, when applied to a symbolic function, the diff command takes the derivative symbolically. For a symbolic function f, diff(f) computes the derivative with respect to x, diff(f,a) computes the derivative with respect to the variable a, and diff(f,2) computes the second derivative of f with respect to x. Likewise, diff(f,a,2) computes the second derivative of f with respect to the symbolic variable a. The command may also be applied to symbolic arrays, for which it returns an array of the same size consisting of derivatives of each element. The following code computes the velocity and acceleration of a particle from its displacement function and then plots the results:

```
EDU>syms x      % define the variable x to be symbolic
EDU>f=exp(x)*sin(pi*x);      % define the function of interest
EDU>fp=diff(f);f2p=diff(f,2);      % compute the first and second derivatives and name them
EDU>ezplot(f,[0,2])
```

The previous code produces the following plot:

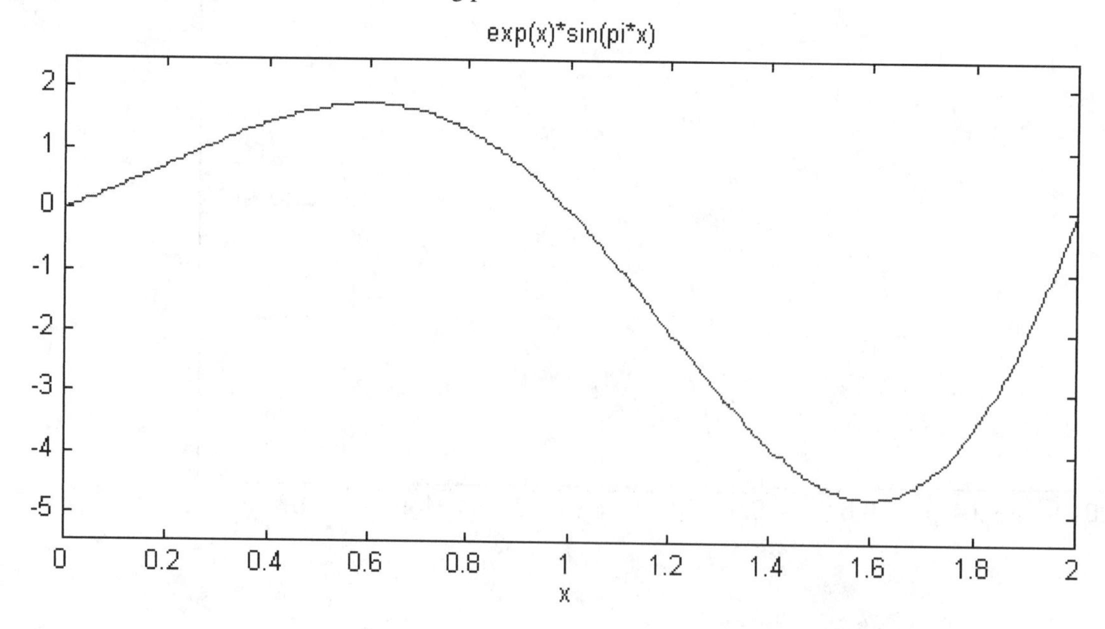

```
EDU>ezplot(fp,[0,2])
```

The previous command produces the following velocity plot:

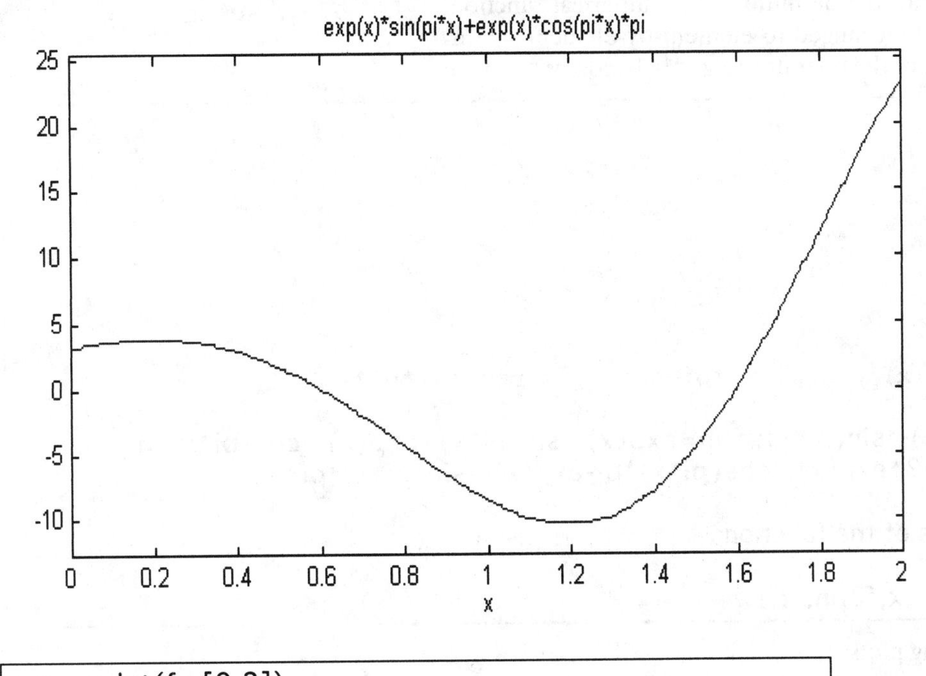


```
EDU>ezplot(fp,[0,2])
```

The previous command produces the acceleration plot:

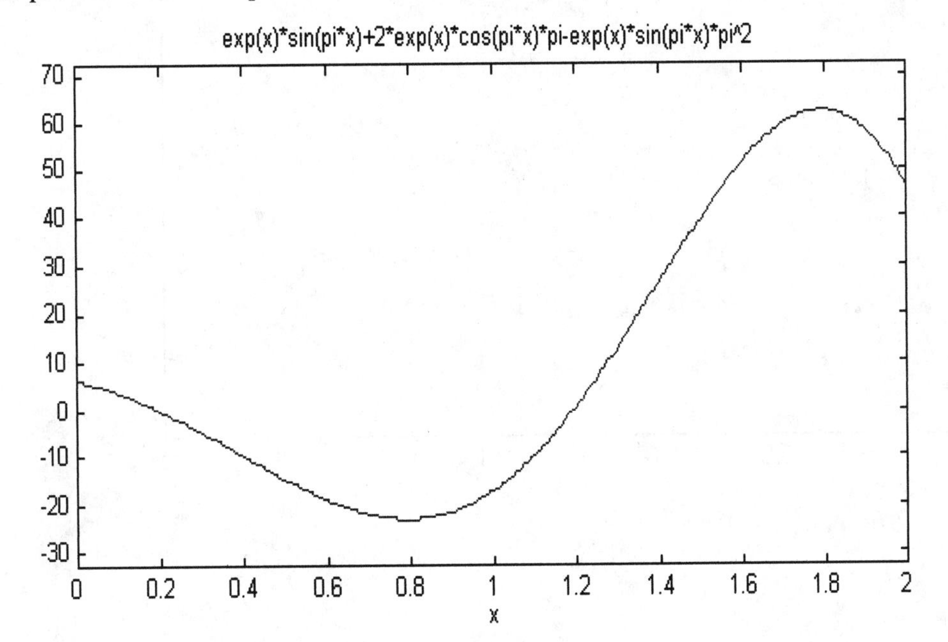

In order to get all three plots on the same graph, the symbolic functions must be converted to numerical functions. This procedure, exhibited in the next set of code, is done by displaying the symbolic derivatives, copying them, and then pasting them into the definition of a numerical function. Then the multiplications must be changed to element-by-element operations by adding periods in front of the operator (e.g., "*" becomes ".*").

```
EDU>f
f =
exp(x)*sin(pi*x)
EDU>fp
fp =
exp(x)*sin(pi*x)+exp(x)*cos(pi*x)*pi
EDU>f2p
f2p =
exp(x)*sin(pi*x)+2*exp(x)*cos(pi*x)*pi-exp(x)*sin(pi*x)*pi^2

EDU>x=(0:.1:2);fn=exp(x).*sin(pi*x); fpn=exp(x).*sin(pi*x)+exp(x).*cos(pi*x)*pi;
f2pn=exp(x).*sin(pi*x)+2*exp(x).*cos(pi*x)*pi-exp(x).*sin(pi*x)*pi^2;

% define numerical versions of the functions

EDU>plot(x,fn,'+',x,fpn,'*',x,f2pn,'x')
```

This code produces the following plots:

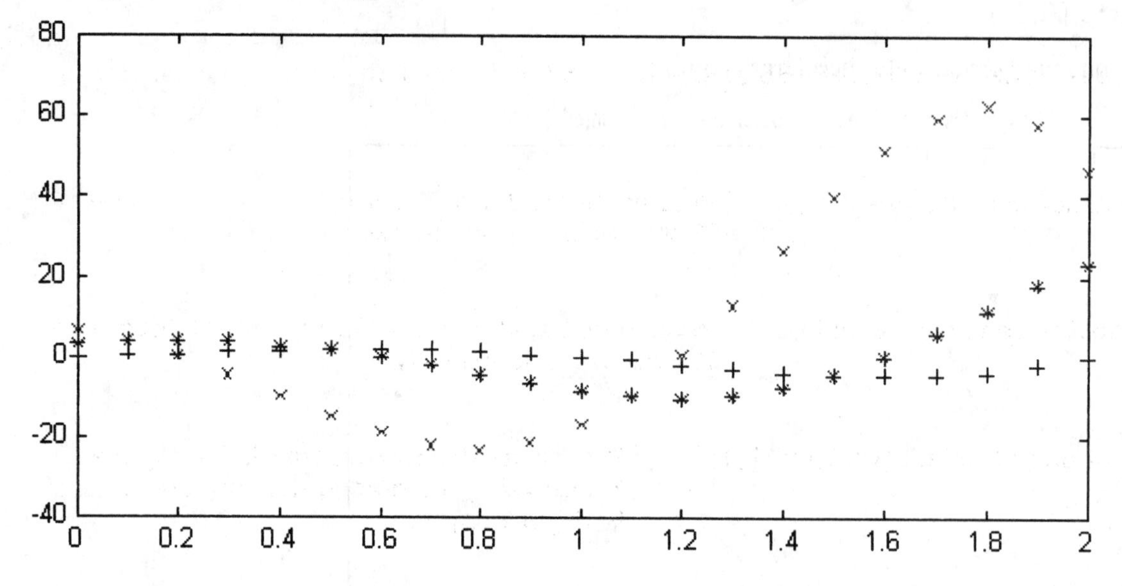

Sample Problem 1.1

An automobile moves along a straight, level section of road such that the displacement is $x(t) = 0.4t^3 + 8t + 10$, where t is in seconds and x is given in feet. Calculate the time it takes the car to reach a speed of 60 mph from its initial state at $t = 0$. How far does the car travel during this time, and what is the value of the acceleration when the car reaches 60 mph?

Notice that the velocity is the derivative of the position (or displacement) and that the displacement will have a local extreme—a maximum or a minimum—when the velocity is zero. In a similar manner, the velocity is a local extreme when the acceleration is zero.

```
EDU>syms t        % define t to be a symbolic variable
EDU>x=0.4*t^3+8*t+10     % define the displacement symbolically
x =
2/5*t^3+8*t+10
EDU>v=diff(x)          % compute the velocity
v =
6/5*t^2+8
EDU>a=diff(v)          % compute the acceleration
a =
12/5*t
EDU>tm=double(solve('6/5*t^2+8=88'))% solve v=88 and display the answer numerically
tm =
    8.1650
   -8.1650
EDU>disp=double(compose(x,tm(1)))      % compute the displacement for the time of interest
                                       % and evaluate it numerically

disp =
  293.0521
EDU>accel=double(compose(a,tm(1)))     % compute the acceleration for the time of interest and
                                       % evaluate it numerically

accel =
   19.5959
EDU>distrav=double(int(abs(v),0,tm(1)))      % compute the distance traveled during the
                                             % interval by integrating the absolute value of
                                             % the velocity

distrav =
  283.0521
```

The previous code uses several of the symbolic operations available on MATLAB. First, the diff command is used to determine the velocity and acceleration from the given displacement. Next, in order to determine the time *tm* when the velocity reaches 88 ft/s, the symbolic command solve is used; this command returns two values, as the equation that is solved is quadratic. The solve command is nested inside the double command, which causes the symbolic result to be evaluated numerically and stored in the vector *tm*. Only the positive root is physical here, so tm(1) is used. The symbolic command compose is used to evaluate the acceleration function and displacement function at the value tm(1). Both of these functions are nested inside of the double command in order to cause numerical values to be returned. The compose command symbolically combines the symbolic function that is its argument, which in this case has the effect of evaluating a symbolic function at a specific numerical value—that is, (tm(1)). To compute the distance traveled, which is different from the value of the displacement, the symbolic integration function is used and then the function is evaluated numerically.

Sample Problem 1.2

During walking, the center of mass of an individual rises and falls, following a sinusoidal motion $y(t) = C\cos(2\pi t - \pi) + y_0$, where y_0 is the height of the center of mass when the individual is standing and C is the amplitude of the displacement of the center of mass. Determine the vertical velocity and acceleration of the center of mass, and graph the change of vertical displacement, velocity, and acceleration for a time of 1s.

The displacement of the center of mass of an individual when walking is given as

$$y(t) = 10\cos(2\pi t - \pi) = 100 \text{ cm}.$$

The velocity and acceleration can easily be determined for this problem by hand, but they are symbolically differentiated here using MATLAB:

```
EDU>syms t
EDU>y=10*cos(2*pi*t-pi)+100;
EDU>v=diff(y)
v =
20*sin(2*pi*t)*pi
EDU>a=diff(v)
a =
40*cos(2*pi*t)*pi^2
EDU>ezplot(y,[0.1])
```

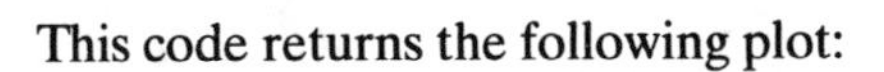

This code returns the following plot:

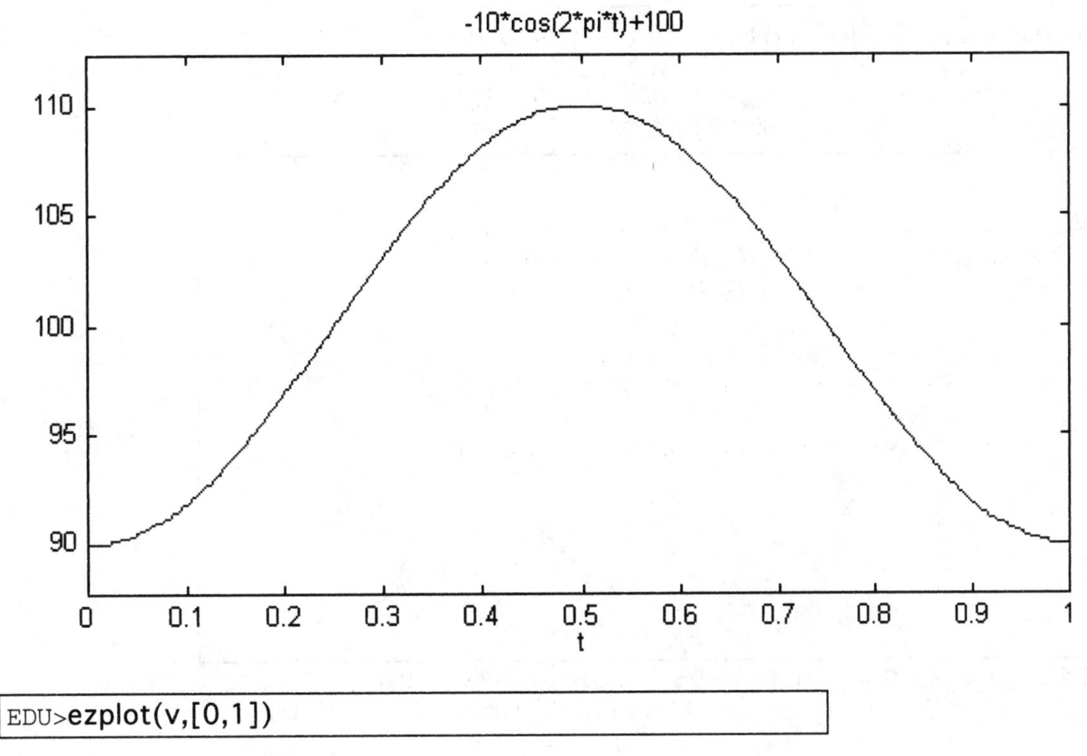


```
EDU>ezplot(v,[0,1])
```

The previous command returns the following plot:

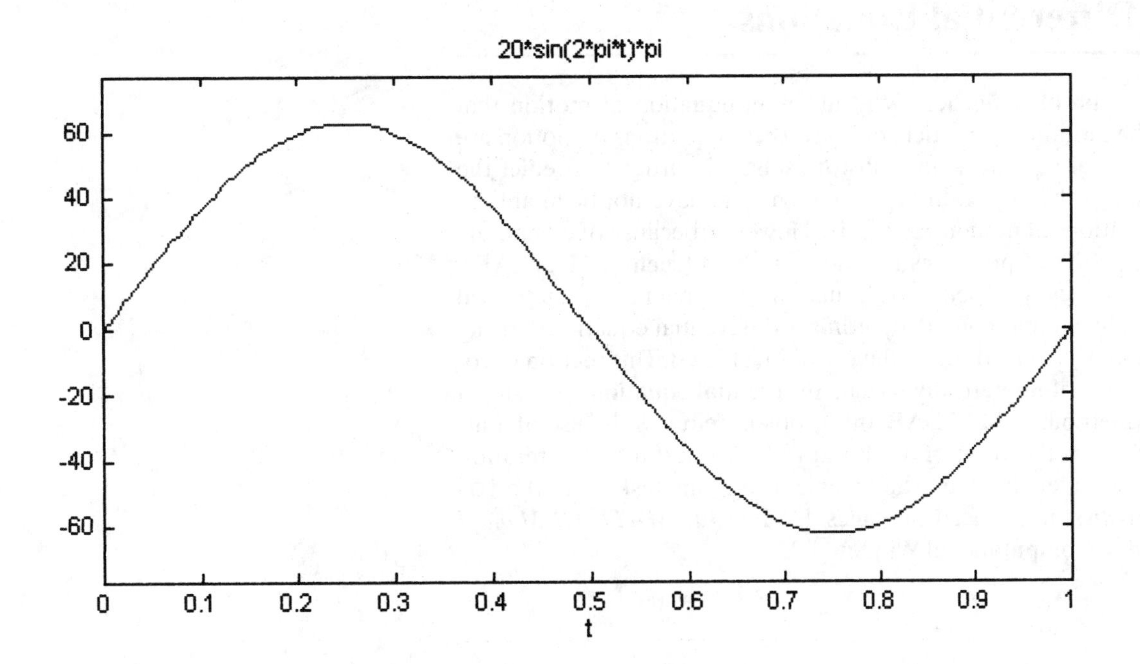

```
EDU>ezplot(a,[0,1])
```

The previous command returns the following plot:

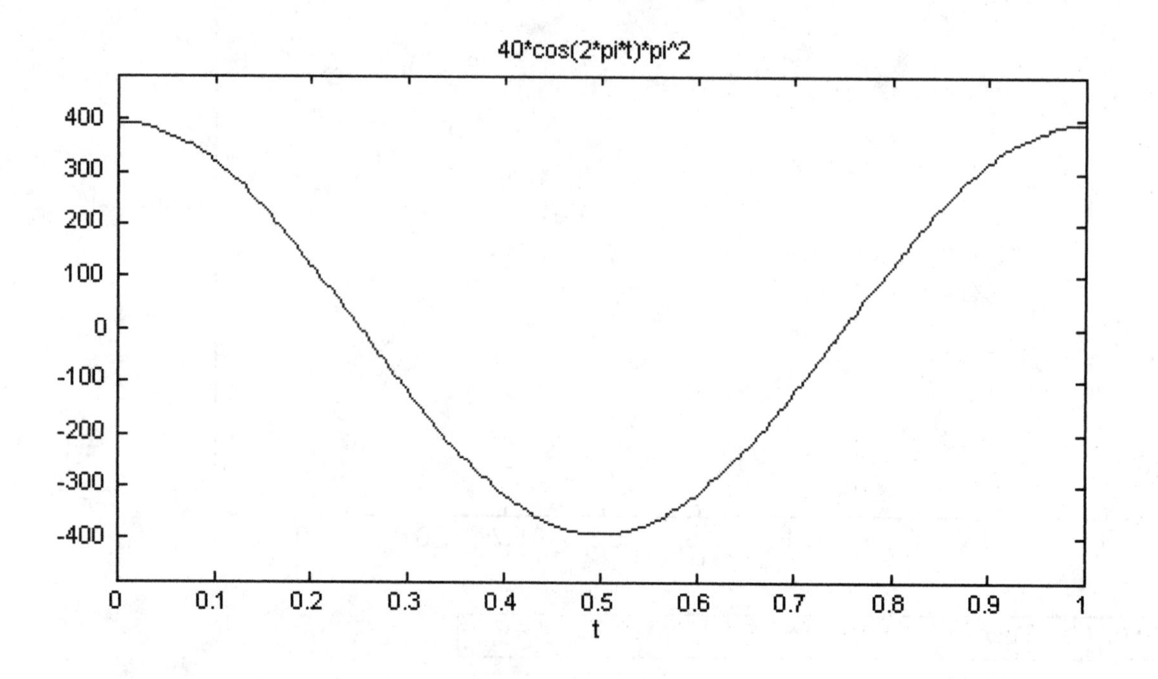

Solving Differential Equations

Most dynamics problems generally result in an equation of motion that determines the motion of a particle or body. These equations of motion are usually differential equations and must be solved in order to predict the path of motion of the object. In the past, engineers have not been able to solve such equations of motion very easily. However, because of computers and the high quality of programs available on them (such as MATLAB), we are now able easily to solve the equations of dynamics by numerical methods. Both linear and nonlinear, ordinary differential equations arising in dynamics can be solved through the use of MATLAB. This section introduces the concept of numerically solving differential equations by using a simple Euler method. In MATLAB, this problem requires the use of a little programming in the form of a loop in order to perform the iteration required in the Euler method. The command for this task—e.g. the `for` command, was first introduced on pages 42-43 of the *MATLAB Manual for Statics* and in Computational Window 4.2.

Consider solving the differential equation

$$\ddot{x}(t) + 4x(t) = 0, \quad x(0) = 0, \; \dot{x}(0) = 1$$

for the interval of time between 0 and 2 seconds. First, we need to write this equation as two first-order equations, rather than one second-order equation. Defining our two new variables to be $x(t)$ and $v(t)$, the previous second-order equation can be written as two coupled first-order equations:

$$\begin{aligned} \dot{x}(t) &= v(t) \\ \dot{v}(t) &= -4x(t) \end{aligned} \quad \text{with initial conditions} \quad \begin{aligned} x(0) &= 0 \\ v(0) &= 1. \end{aligned}$$

The MATLAB code for solving this particular problem using the Euler method is given in the following m-file, which is saved as `euler.m`.

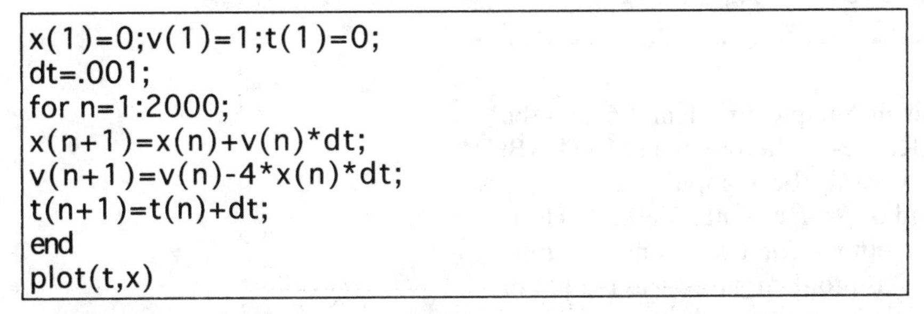

```
x(1)=0;v(1)=1;t(1)=0;
dt=.001;
for n=1:2000;
x(n+1)=x(n)+v(n)*dt;
v(n+1)=v(n)-4*x(n)*dt;
t(n+1)=t(n)+dt;
end
plot(t,x)
```

Typing `euler` after the prompt then produces the following plot of the time history of the motion of the particle:

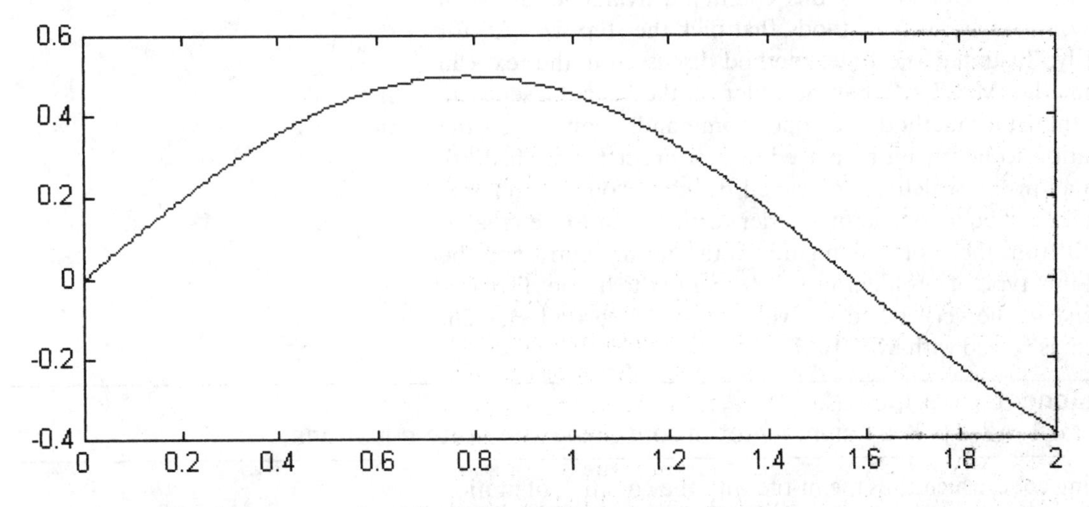

As suggested in the text, you should use this m-file to change dt to larger and smaller values, and note the change on the computed response. In this case, the differential equation is linear and has an analytical solution that may be used for comparison with the numerical solution produced by the Euler method. If you want to change the length of time over which the solution is computed, you must change both *dt* and the last element (element 2000) in the `for` command. In MATLAB, it is more convenient to use a Runge–Kutta method for solving differential equations, as this method is more accurate and computes the required *dt* automatically. The use of the Runge–Kutta formulation in MATLAB is presented in the next example.

Sample Problem 1.6

Determine the acceleration, velocity, and displacement as functions of time if the acceleration is given as

$$a(x) = 3x^2 - 8 \text{ ft/s}^2$$

and the initial position is $x_0 = 0$. The initial velocity is $v_0 = 2$ ft/s.

The nonlinear differential equation in Sample Problem 1.6 may be solved by numerical integration using a Runge–Kutta routine in MATLAB. The time response of any system may easily be computed by simple numerical means such as Euler's method or Runge–Kutta methods. Here, the use of these common numerical methods for computing the time response using MATLAB is presented. This problem introduces the use of numerical methods for solving differential equations.

In MATLAB, the main command for solving differential equations is "ode". There are several versions of this command available, many of which are based on Runge–Kutta methods, that pick the step size (Δt) for you. As previously illustrated, the Euler method discussed in the text can also be programmed in MATLAB, but the Euler Method is not a separate command. Like the Euler method, the "ode" commands require that the differential equation to be solved be placed in first-order form. This task must be done in an m-file, which is then called from the "ode" command. The m-file contains the equations in first-order form and must be created before it is called from the "ode" command. Note that in contrast to the development of this type of problem in the *Dynamics* text, the displacement is listed first in the vector and the velocity is the second element when the problem is solved with MATLAB.

```
function xdot=oneptsix(t,x);
xdot=[x(2); 3*x(1)^2-8]; % a column vector containing the state equations
```

Next, the following code, which calls the m-file with the equation of motion and then plots it, is typed in the command window:

```
EDU>tspan=[0 4];     % states the range of values for time
EDU>x0=[0 2];        % enters the initial conditions
EDU>ode45('oneptsix',tspan,x0)
```

Running the previous set of commands produces the following response:

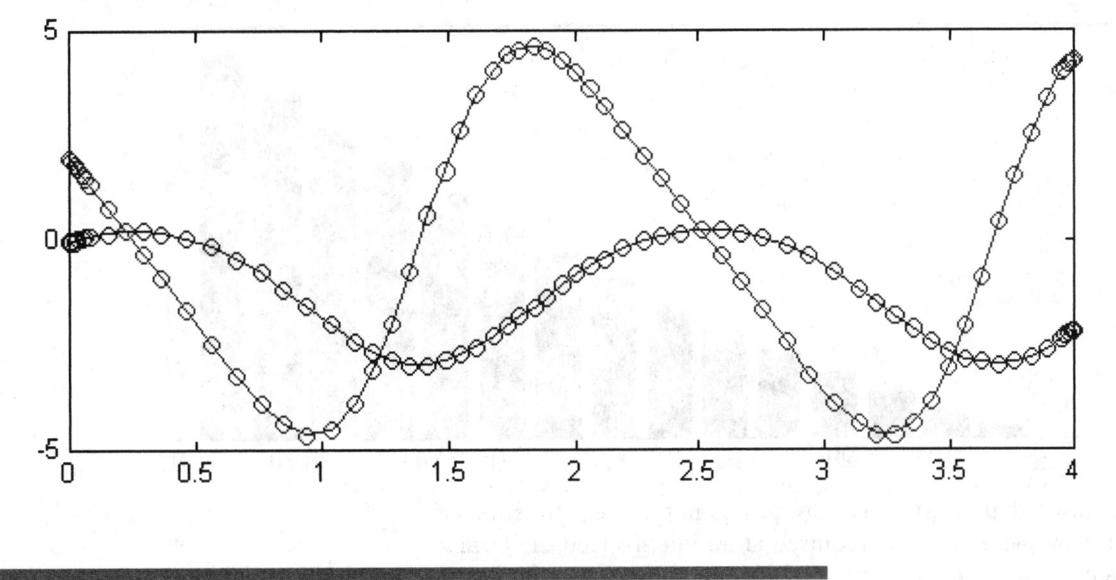

Sample Problem 1.7

During a "locked-up" skid, a car has a constant negative acceleration (braking deceleration). If there is a 1.75-second reaction time before braking (the average time observed for drivers), and the braking deceleration is 22.5 ft/s^2, determine the distance to stop if the initial speed is (a) 60 mph; (b) 45 mph; and (c) 30 mph.

The results of Sample Problem 1.7 can be graphed for easier interpertation or for inclusion in a report or manual. The commands for this task include MATLAB'S "bar" command:

```
EDU>s=(0:5:75);
EDU>v0=1.467*s;
EDU>xf=1.75*v0+(1/(2*22.5))*v0.^2;
EDU>bar(s,xf)
```

The previous set of commands produces the following bar chart:

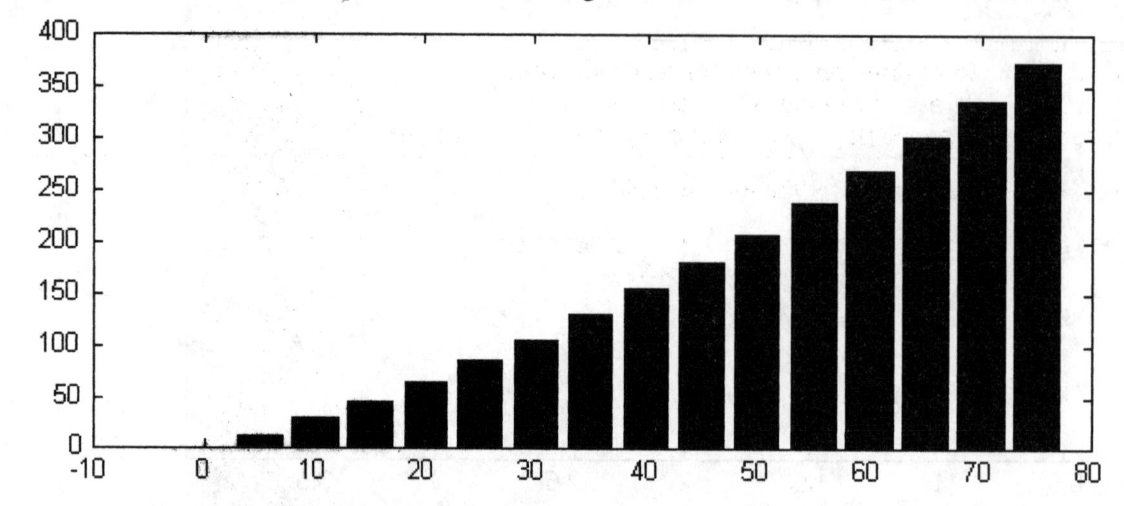

The bar chart shows that the stopping distance is not a linear function of the speed. For example, 120 feet is required at an initial speed of 30 mph and 326 feet is needed at an initial speed of 60 mph. Bar charts of this type of information are included in most driver information manuals. This data is based upon stopping distances on dry pavement.

Sample Problem 1.10

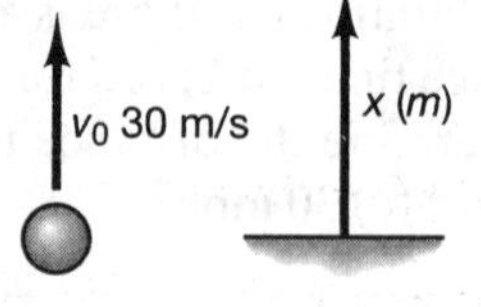

A ball is thrown upward against gravitational attraction and air resistance proportional to the square of the velocity. Air resistance always opposes the motion; that is, it has the opposite sign of the velocity. The acceleration can be specified as $a(v) = -g - cv^2 \text{ sign}(v)$, where $c = 0.001$ (1/m) and $g = 9.81 \text{ m/s}^2$. If the ball is thrown straight up with an initial velocity of 30 m/s, determine the velocity–displacement relationship.

Find the maximum height the ball reaches with and without air resistance.

The differential equation of motion is a nonlinear function of v and contains the absolute value of v. This equation can be numerically integrated, as illustrated in the next set of boxes. As in previous examples, an m-file must first be established to define the equations of motion.

```
function xdot=oneten(t,x);
xdot=[x(2);-9.81-0.001*x(2)*abs(x(2))];     % define the equation of motion
```

The m-file is then called by entering the following commands in the command window:

```
EDU>tspan=[0 6];      % define the time interval of interest
EDU>x0=[0;30];        % enter the initial conditions
EDU>ode45('oneten',tspan,x0);      % call the Runge-Kutta routine
```

The previous set of commands produces the following plot:

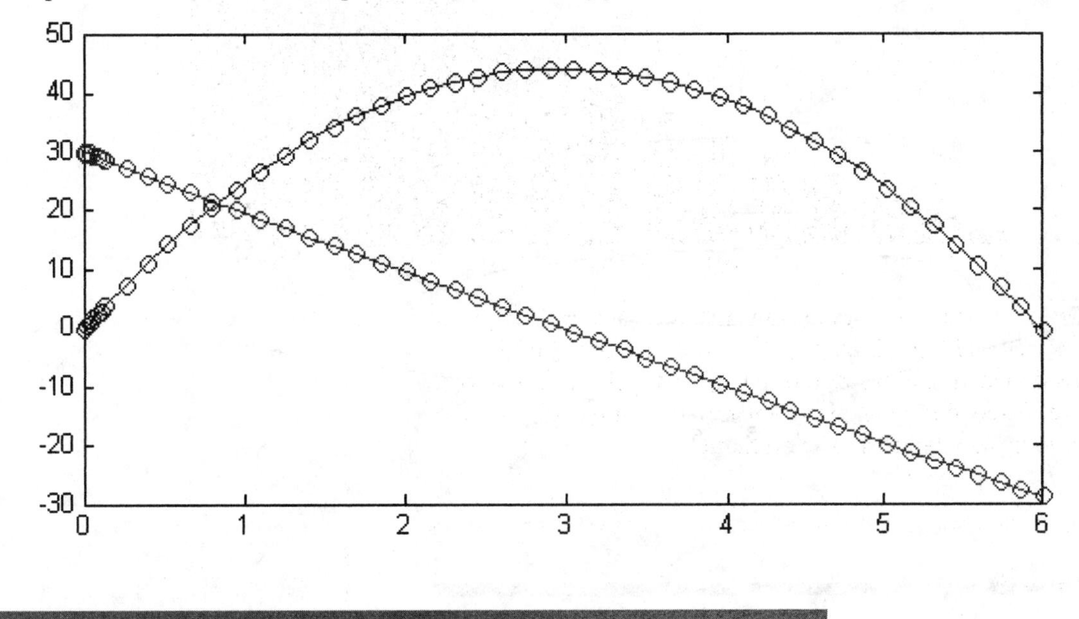

Sample Problem 1.12

Consider a particle moving in a helical motion at a constant pitch. The position vector can be given as

$$\mathbf{r}(t) = (R\cos\omega t)\hat{\mathbf{i}} + (R\sin\omega t)\hat{\mathbf{j}} + (pt)\hat{\mathbf{k}},$$

where R is the radius of the helix, p is the pitch of the helical curve in space, and ω determines the time to complete one cycle around the helix. Determine the velocity and acceleration of the particle at any time in terms of the constants R, p, and ω.

Computational software is certainly not required for solving this problem. However, plotting the response in three dimensions provides a visualization of the path of motion. Here, 40 points are used to generate the helix. The radius is chosen to be 5, and the pitch is chosen to be 0.2. It takes 20 seconds to complete one cycle around the helix. Three-dimensional plots are obtained by the command `plot3`.

```
EDU>i=(0:1:40);
EDU>x=5*cos(0.1*i*pi);
EDU>y=5*sin(0.1*i*pi);
EDU>z=0.2*i;
EDU>plot3(x,y,z),xlabel('x'),ylabel('y'),zlabel('z')
```

This set of commands yields the following plot:

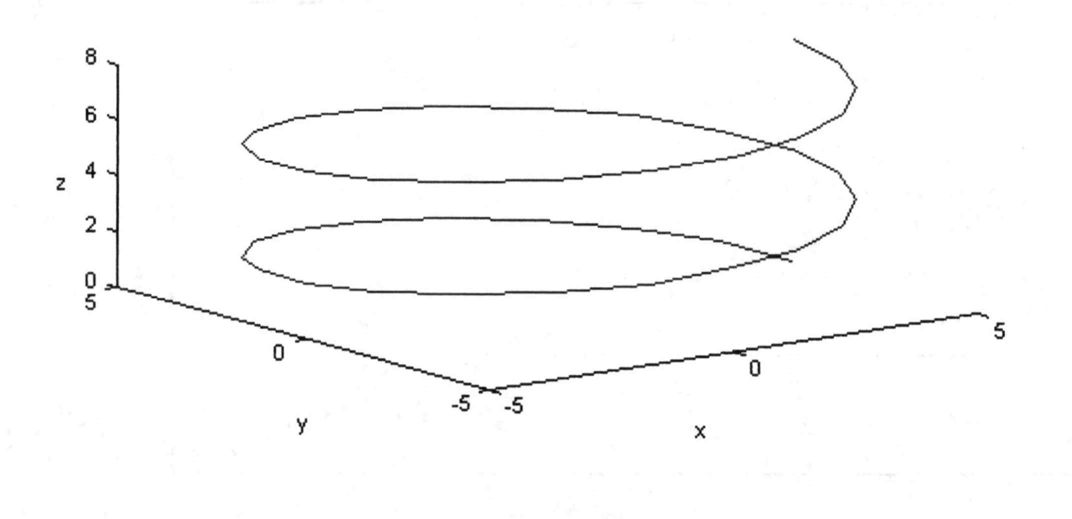

Sample Problem 1.14

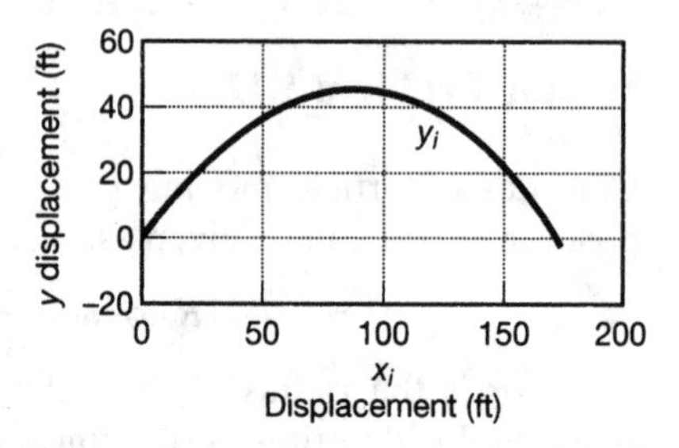

The punter kicks the football into a heavy rain such that the air resists the motion in a manner proportional to the velocity squared. If the initial velocity of the ball is 80 ft/s and the ball leaves the punter's foot at a 45° angle, determine the distance the ball travels. The acceleration is $\mathbf{a} = -g\hat{\mathbf{j}} - c\mathbf{v}|\mathbf{v}|$, where $c = 0.001(\text{ft}^{-1})$.

In this problem, the vector differential equation of motion is nonlinear and the two scalar equations are coupled. In Sample Problem 1.13, the vector differential equation was linear, the component equations were uncoupled, and computational software was not used to compute the solution. However, both problems can be solved numerically with the same basic code by setting the damping term to zero in Sample Problem 1.14 to obtain the equation of Sample Problem 1.13. Here, the equations of motion are first put into first-order form, with the notation of $x(1)=x$, $x(2)=v_x$, $x(3)=y$, and $x(4)=v_y$, and then are placed in the m-file named "fourteen.m":

```
function xdot=fourteen(t,x);
c=0.001;g=32.2;
xdot=[x(2);-c*x(2)*sqrt(x(2)^2+x(4)^2);x(4);-g-c*x(4)*sqrt(x(2)^2+x(4)^2)];
```

The next set of commands enters the time interval of interest and the initial conditions and then calls the equations of motion contained in "fourteen.m". Rather then plotting the response, as we did in the previous sample problem, we write the response to the matrix **x**. The notation "x(:,1)" refers to the first column in this matrix, which contains the *x* displacement, and `x(:,3)` refers to the third column in the matrix, which contains the *y* displacement.

```
EDU>tspan=[0 3.4];
EDU>x0=80*[0;cos(45*pi/180);0;sin(45*pi/180)];
EDU>[t,x]=ode45('fourteen',tspan,x0);
EDU>plot(x(:,1),x(:,3))
```

The previous command window produces the following plot of `y` versus `x`, indicating the trajectory:

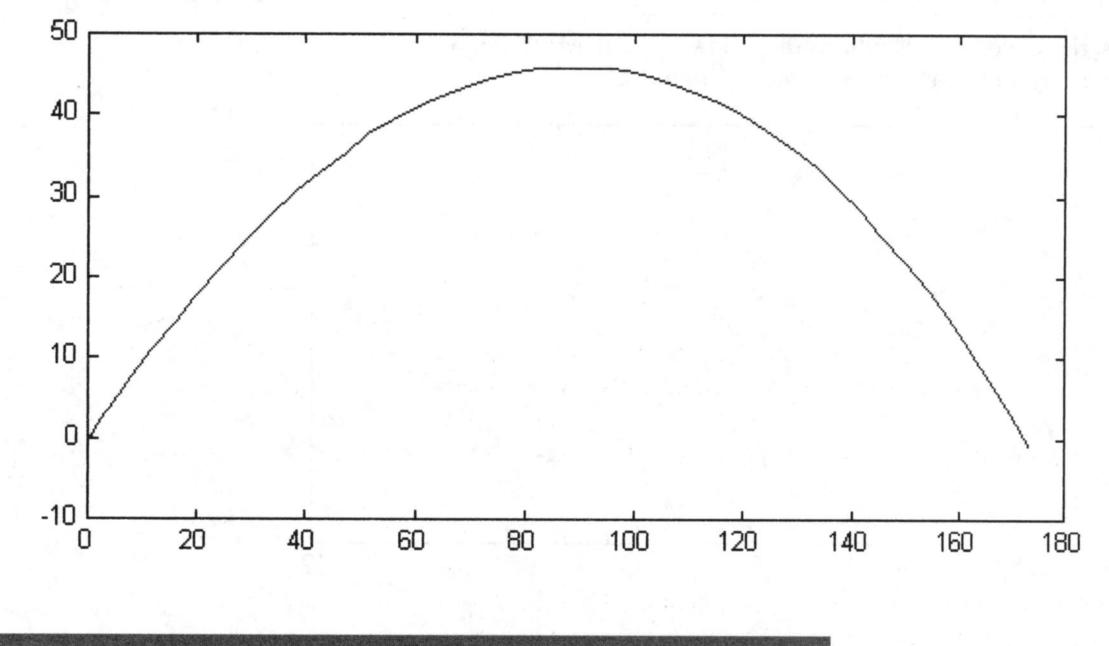

Sample Problem 1.15

A car, starting from rest, drives around a curve with a radius of 1000 feet. If the car undergoes a constant tangential acceleration of 8 ft/s^2, determine the total acceleration after 10 seconds.

The computations in Sample Problem 1.15 are straightforward; however, the use of MATLAB allows for the visualization of the components and magnitude of the acceleration vector simply by plotting their values versus time. First, t, the row vector of time values, is created. Then the radius of curvature ρ is entered and the length r of the vector t is computed. Next, the matrix a is formed; a consists of a first row that contains the value 8 in each of the r entries (entered with the command "8*ones(1,r)") and a second row that contains the values for $16t^2/\rho$. Next, the magnitude of the acceleration vector at each instant of time must be computed. Here, "a(1,:).^2" computes the square of each element of the first row of a, and "a(2,:).^2" computes the square of each element of the second row of a. All of these squares are added together and the square root of the sum is taken to form a 1-x-r vector containing the magnitude at each time interval.

```
EDU>t=(0:0.5:12);
EDU>p=1000;r=length(t);
EDU>a=[8*ones(1,r);(64*t.^2)/p];
EDU>mag=sqrt(a(1,:).^2+a(2,:).^2);
EDU>plot(t,mag,'+',t,a(1,:),'*',t,a(2,:),'x')
```

The previous series of commands returns the following plots, which can be labeled, if desired, by adding a "plot" statement with titles:

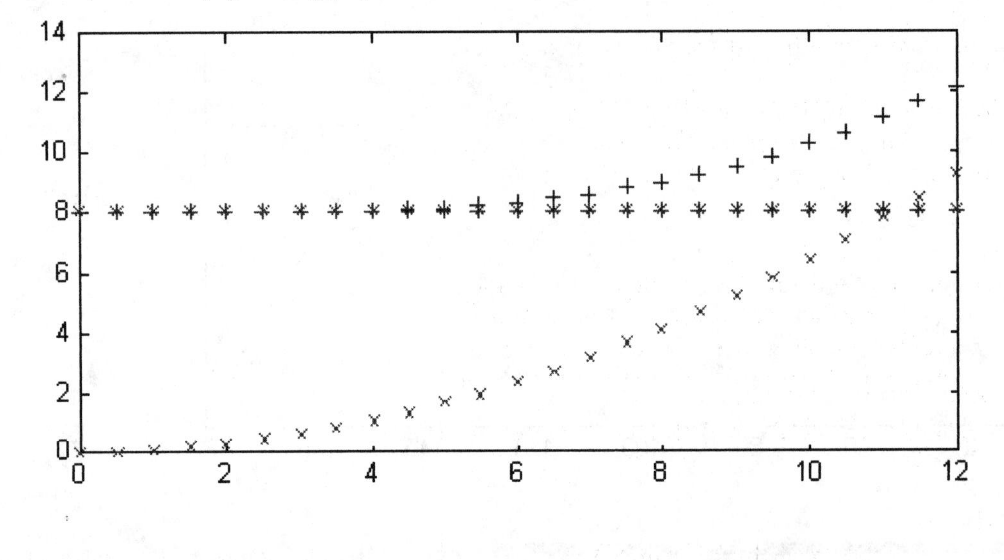

Sample Problem 1.16

A particle moving along a circular path undergoes an angular acceleration given by $\alpha = -2\omega + 3t^2$ rad/s^2. If the particle starts at rest, determine the angular velocity and the angular postion as functions of time.

SOLUTION:

The angular acceleration can be written as the linear first-order differential equation in Eq. (1.91):

$$\frac{d\omega}{dt} + 2\omega = 3t^2,$$

subject to the initial condition $\omega(0) = \omega_0 = 0$. The integrating factor is $\lambda(t) = e^{\int 2dt} = e^{2t}$. Therefore, the general solution is

$$\omega(t) = \frac{\int 3t^2 e^{2t} dt + C}{e^{2t}} = \frac{3}{2}\left(t^2 - t + \frac{1}{2}\right) - Ce^{-2t}.$$

Since the angular velocity is zero at $t = 0$, the constant is $C = -3/4$. Hence, the angular velocity is

$$\omega(t) = \frac{3}{2}\left(t^2 - t + \frac{1}{2}\right) - \frac{3}{4}e^{-2t}.$$

The angular postion (in radians) can be obtained by integrating $\omega(t)$. Doing so, we obtain

$$\theta(t) = \frac{t^3}{2} - \frac{3t^2}{4} + \frac{t}{2} + \frac{3}{8}e^{-2t} + C_2,$$

where C_2 is a constant of integration. If $\theta(0) = 0$, the value of this constant is $C_2 = -3/8$.

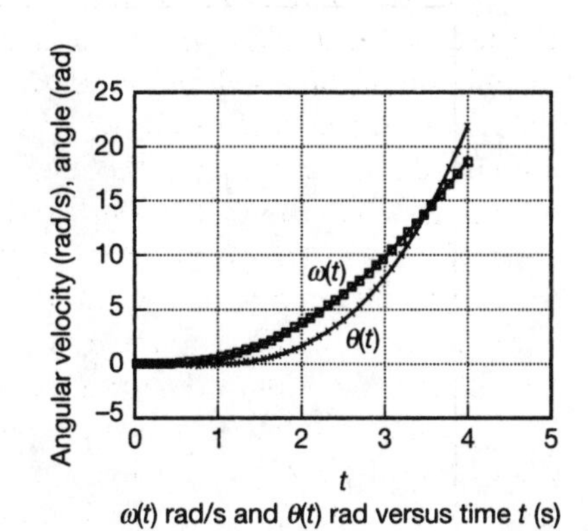

$\omega(t)$ rad/s and $\theta(t)$ rad versus time t (s)

If the radius of the circular path in Sample Problem 1.16 is 1.2 meters, determine general functions of time for the tangential and normal acceleration and the magnitude of the total acceleration. Plot the results.

The numerical calculations required for this problem are similar to those of the previous sample problem; the elements of an acceleration vector and its magnitude are computed and plotted as a function of time.

```
EDU>r=1.2;t=(0:0.1:2);l=length(t);
EDU>w=(3/2)*(t.^2-t+.5)-.74*exp(-2*t);
EDU>a=3*(t-.5)+1.5*exp(-2*t);
EDU>at=r*a;an=r*w.^2;
EDU>mag=sqrt(at.^2+an.^2);
EDU>plot(t,mag,'+',t,at,'*',t,an,'x'),xlabel('time')
```

Running this set of commands produces the next plot, which shows the normal and tangential acceleration as well as the magnitude of the acceleration vector, all as functions of time during the first two seconds. Grids and other labels may be added to the plot as desired. Again note that it is important to think of time histories in terms vectors of numbers when working in MATLAB.

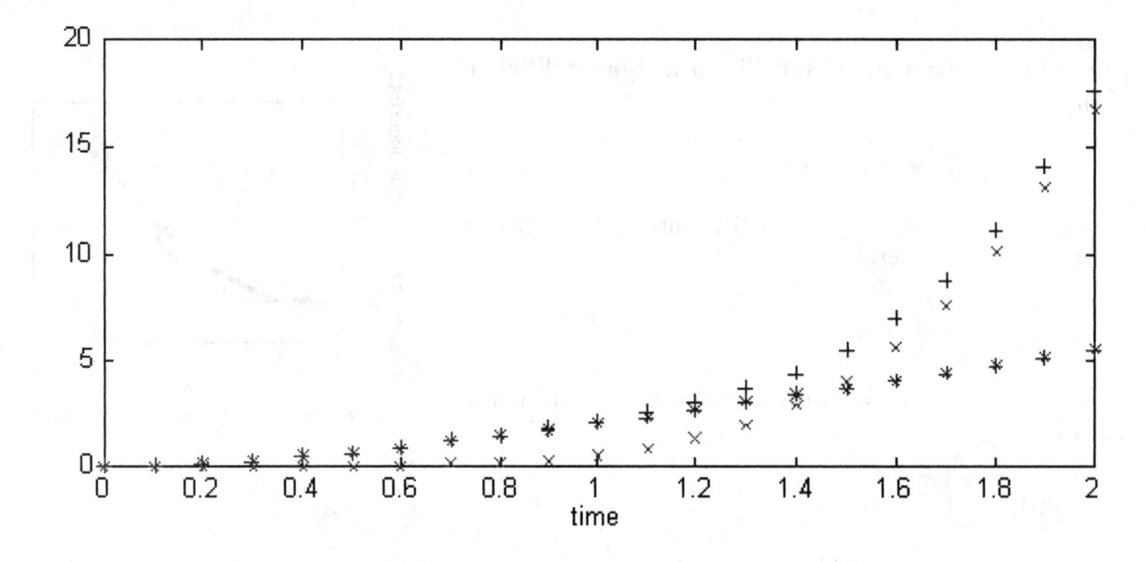

Sample Problem 1.18

Consider particle's position vector, given in meters:

$$\mathbf{r}(t) = t^2\mathbf{i} + 3t\mathbf{j} - t^3\mathbf{k}.$$

Determine the base tangential, principal normal, and binormal vectors, and the radius of curvature when time equals 2 seconds.

For this problem, we make plot similar to the one we constructed for Sample Problem 1.17, but in this case, all the calculations are made symbolically and are thus a little easier to follow:

```
EDU>syms t                      % define t as a symbolic variable
% define the position, velocity and acceleration vectors
EDU>r=[t^2;3*t;-t^3];v=[2*t;3;-3*t^2];a=[2;0;-6*t];
EDU>et=v/sqrt(v'*v);      % compute a unit vector along v(t)
EDU>at=dot(a,et)*et;       % compute the tangential acceleration
EDU>an=a-at;                    % compute the normal acceleration
EDU>p=v'*v/sqrt(an'*an);
EDU>ezplot(p,[0 3]),title('radius of curvature vs time')
```

The previous series of commands produces the analytical expressions for the vectors for the components of the acceleration and for the curvature as functions of time. To find a numerical value for each variable, use the compose command as follows:

```
EDU>double(compose(p,2))
ans =
 50.2969
```

The "plot" command in the first box of code for this sample problem produces the following plot:

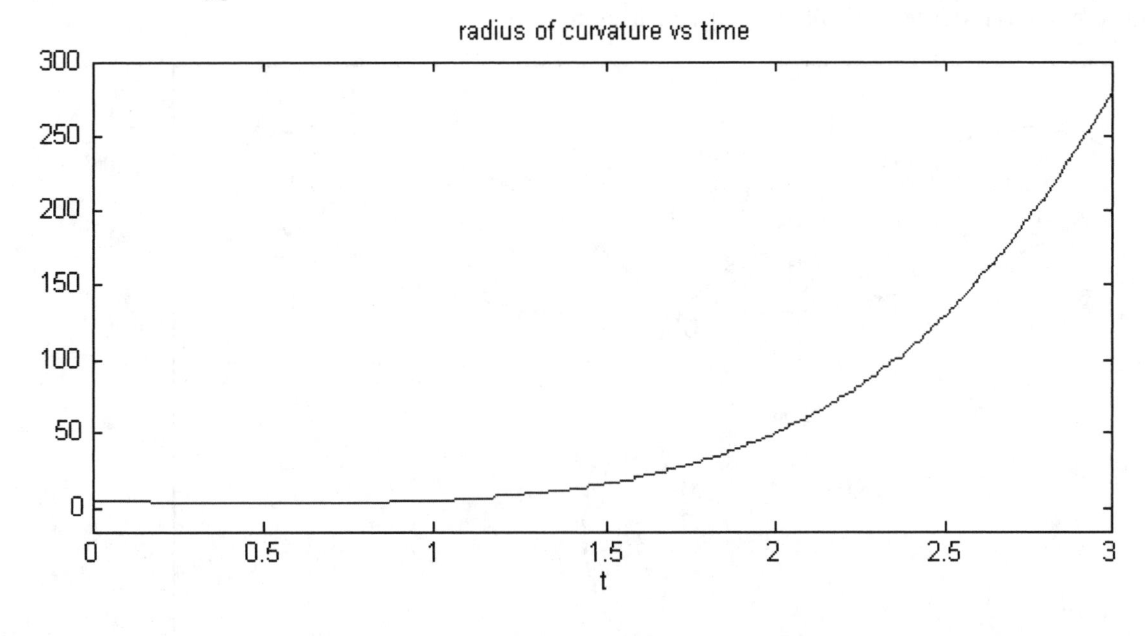

Sample Problem 1.19

Consider a particle's path of motion, given by

$$\theta(t) = \pi t \text{ rad}$$

$$r(t) = 2 \sin 3\theta(t) \text{ m.}$$

Plot the path of motion, and determine the velocity and acceleration of the particle.

The position vector is given in polar coordinates, and to help us visualize the motion, a graph has been created for the position in the *xy* plane. In addition, the components of the acceleration have been calculated and plotted.

```
EDU>t=linspace(0,2);          % generates 100 values between 0 and 2
EDU>r=2*sin(3*pi*t);
EDU>x=r.*cos(pi*t);
EDU>y=r.*sin(pi*t);
EDU>plot(x,y)
```

This series of commands produces the following three-leaved rose:

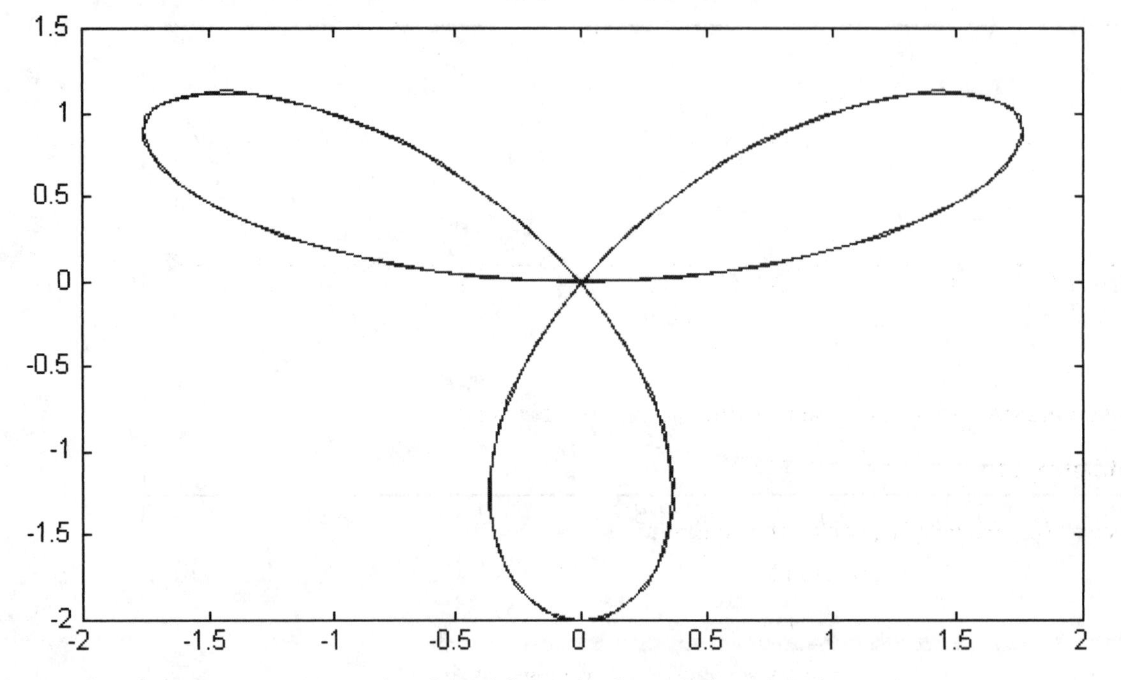

The acceleration components may be plotted from the formulation given in the text of the sample problem by using the following commands:

```
EDU>t=linspace(0,2);
EDU>ar=-(20*pi^2)*sin(3*pi*t);        % the radial component
EDU>ath=(12*pi^2)*cos(pi*t);          % the theta component
EDU>plot(t,ar,'+',t,ath,'*'),xlabel('time'),title('acceleration components')    % plot and label
```

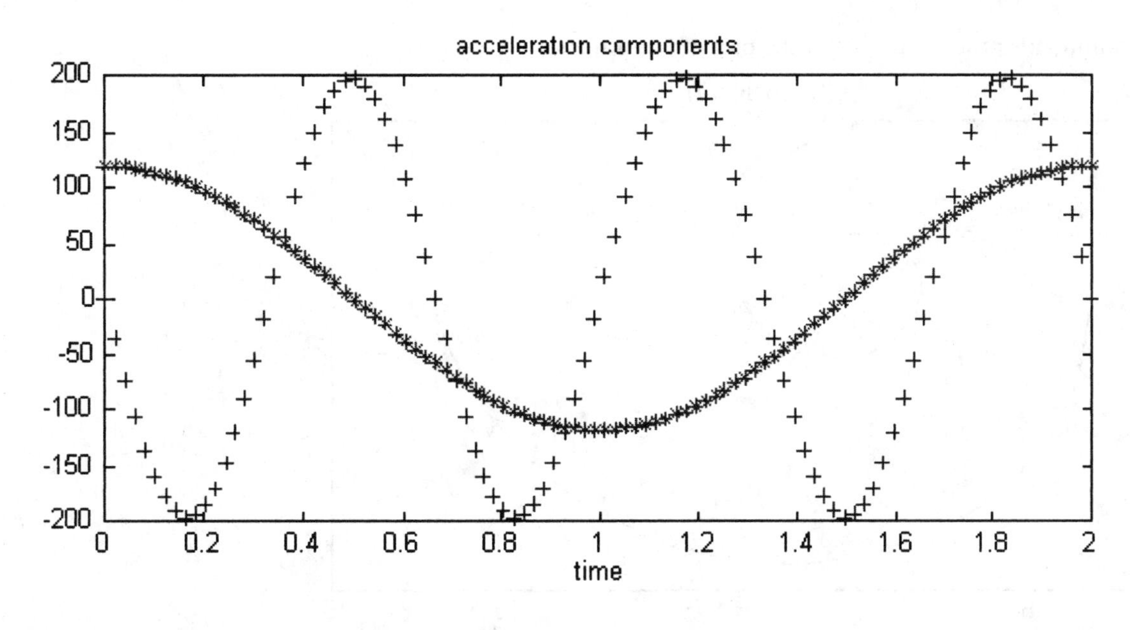

Sample Problem 1.20

A spiral trajectory of a particle is given by the equations

$$\theta(t) = \pi t \text{ rad}$$

$$r(t) = e^{0.2\theta} \text{ m.}$$

Plot the trajectory in space, and determine the r- and θ-components of the acceleration.

This problem is similar to the previous sample problem, and just requires plotting the y components versus the x components of the motion in order to visualize the path of the particle.

```
EDU>t=linspace(0,4);
EDU>th=t*pi;
EDU>r=exp(0.2*th);x=r.*cos(th);y=r.*sin(th);
EDU>plot(x,y),title('y(t) versus x(t)')
```

The previous commands produce the following plot of the trajectory:

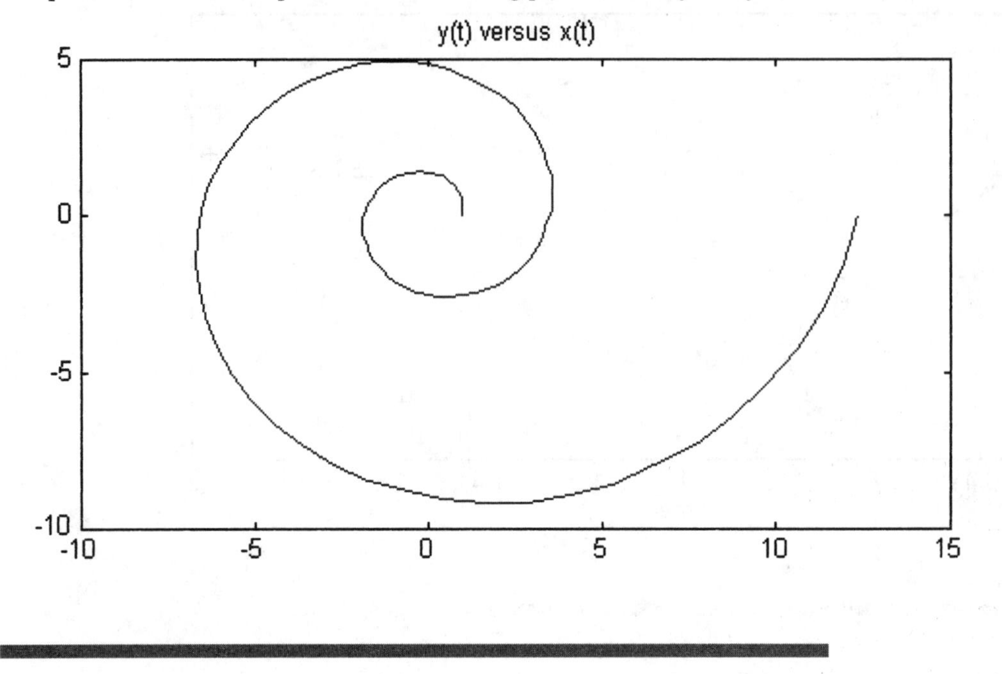

Sample Problem 1.22

Determine the path of motion of a particle if its acceleration is $a_r = 0.5$ m/s^2 and $a_\theta = 2$ m/s^2, and its initial velocity is $v_r = 1$ m/s and $v_\theta = 0.5$ m/s. At time equal to zero, the particle has initial coordinates $r = 0.2$ m and $\theta = 0$.

The MATLAB code we will use to solve this problem is the same program we used to solve Sample Problem 1.14, which also involves two coupled differential equations. Using that code, the following m-file is prepared to contain the equations of motion in first-order form, with the accelerations entered second (note that this order is the opposite that given in the text):

```
function xdot=twotwo(t,x);
xdot=[x(2);0.5*x(1)*x(4)^2;x(4);2-2*x(2)*x(4)/x(1)];
```

Here, the variables in first-order form are $x(1) = r_0, x(2) = vr_0, x(3) = \theta_0$, and $x(4) = \omega_0$, so that the initial conditions become $x_0 = [\ 0.2\ 1.0\ 0\ 2.5]$ and the first-order form of the equations is

$$\begin{bmatrix} \dot{r}(t) \\ \dot{v}(t) \\ \theta(t) \\ \omega(t) \end{bmatrix} = \begin{bmatrix} v(t) \\ 0.5 + r(t)\omega^2(t) \\ \omega(t) \\ 2 - 2\frac{2v(t)\omega(t)}{r(t)} \end{bmatrix}.$$

Once the solution to this first-order differential equation is obtained, the trajectory can be visualized in rectangular coordinates by using the following transformation:

$$x(t) = r\cos(\theta) \text{ and } y(t) = r\sin(\theta).$$

The solution is computed, transformed to rectangular coordinates, and plotted by the following series of commands:

```
EDU>tspan=[0 2];
EDU>x0=[0.2;1.0;0;2.5];
EDU>[t,x]=ode45('twotwo',tspan,x0);
EDU>xn=x(:,1).*cos(x(:,3));yn=x(:,1).*sin(x(:,3));
EDU>plot(xn,yn),title('x-y trajectory')
```

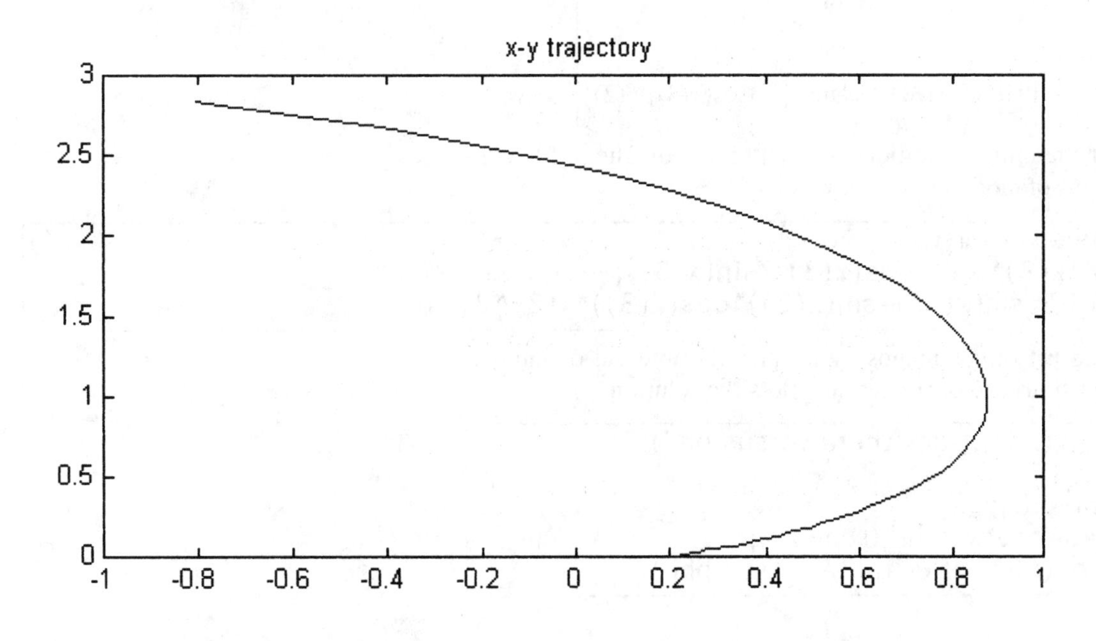

Sample Problem 1.24

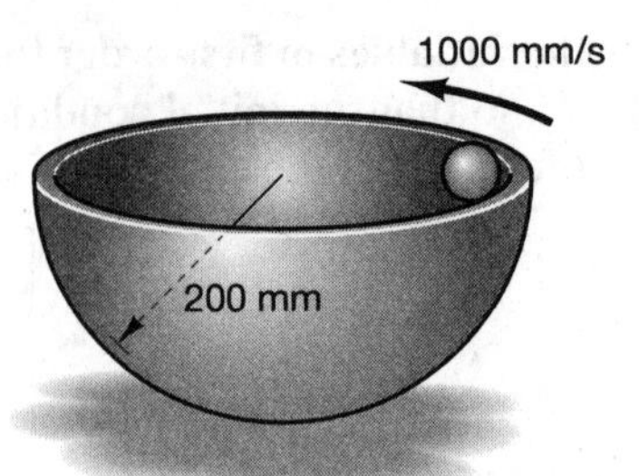

A particle is released at the rim of a smooth hemispherical bowl of radius 200 mm with an initial velocity in the circumferential direction of $R\dot{\theta}$ = 1000 mm/s, and it slides into the bowl under the influence of gravity (g = 9810 mm/s^2). Write the kinematic equations of motion, and solve the resulting nonlinear differential equations using computational software.

This sample problem involves two coupled differential equations that need to be solved together using a four-component vector differential equation, as in the previous sample problem. The equations of motion are put in first-order form with displacements given first, followed by the corresponding velocity. (Note that the order is the opposite of that given in the *Dynamics* text.) Here, the first-order variables are named as follows:

$$x(1) = \theta_0, x(2) = \dot{\theta}_0, x(3) = \phi_0, \text{and } x(4) = \dot{\phi}_0.$$

With this notation, the equations of motion in first-order form become

$$\begin{bmatrix} \dot{x}(1) \\ \dot{x}(2) \\ \dot{x}(3) \\ \dot{x}(4) \end{bmatrix} = \begin{bmatrix} x(2) \\ -2x(2)x(4)\dfrac{\cos(x(3))}{\sin(x(3))} \\ x(4) \\ 49.05\sin(x(3)) + \sin(x(3))\cos(x(3))x^2(2) \end{bmatrix},$$

and the vector for the initial conditions is $x_0 = [0\ 5\ \pi/2\ 0]$. The m-file containing the equations of motion is as follows:

```
function xdot=onetwofour(t,x);
xdot=[x(2);-2*x(2)*x(4)*cos(x(3))/sin(x(3));
      x(4);49.05*sin(x(3))+sin(x(3))*cos(x(3))*x(2)^2];
```

Next, the following list of commands establishes the interval of interest and the initial conditions and computes and plots the solution:

```
EDU>plot(x(:,1),x(:,3)),title('theta versus phi')
EDU>tspan=[0 0.8];
EDU>x0=[0;5;pi/2;0];
EDU>[t,x]=ode45('onetwofour',tspan,x0);
EDU>plot(x(:,1),x(:,3)),title('theta versus phi')
```

This series of commands produces the following plot of the trajectory:

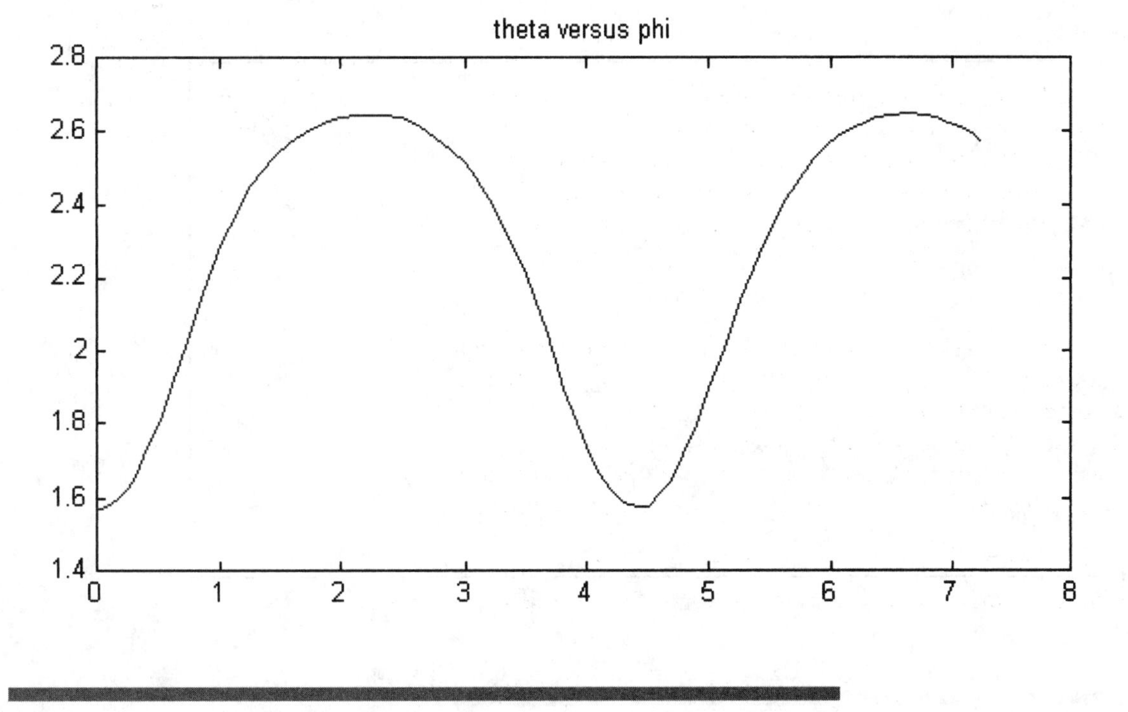

Sample Problem 1.25

Two cars begin moving from rest at the same time, but car *A* is 100 ft behind car *B* at the start. Car *A* maintains a constant acceleration of 8 ft/s^2, and car *B* has a constant acceleration of 6 ft/s^2. How much time is required for car *A* to overtake car *B*, and how much distance will car *A* have traveled when it overtakes car *B*?

The relative motion between two cars can be computed and then used to create a plot of the absolute motion of each car during the period of time of interest. This computation amounts to using MATLAB code as a graphical calculator.

```
EDU>t=(0:0.5:15);
EDU>xa=4*t.^2;xb=3*t.^2+100;
EDU>plot(t,xa,'+',t,xb,'*'),title('xa,+, and xb, *, versus t')
```

The previous set of commands yields the following plot:

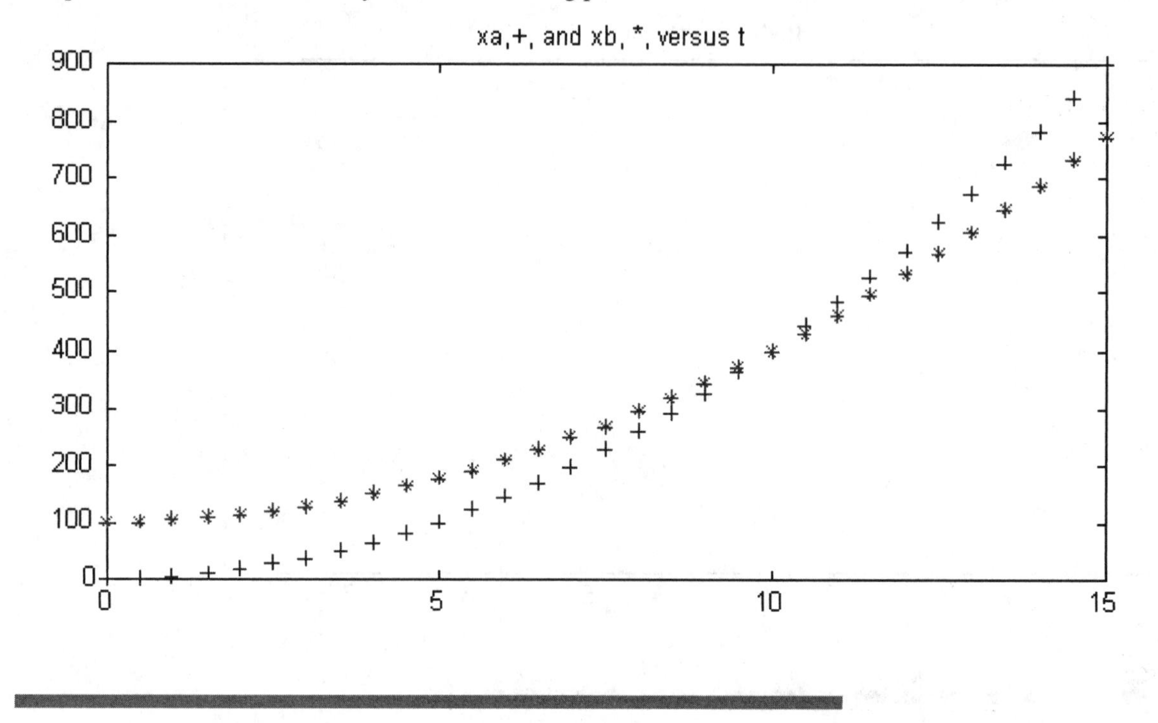

Sample Problem 1.26

A surfer paddles her board out against the incoming waves; the constant velocity of her board relative to the water is $v_{B/W}$. The velocity of the waves is sinusoidal with time and decreases with distance from the beach according to the relationship

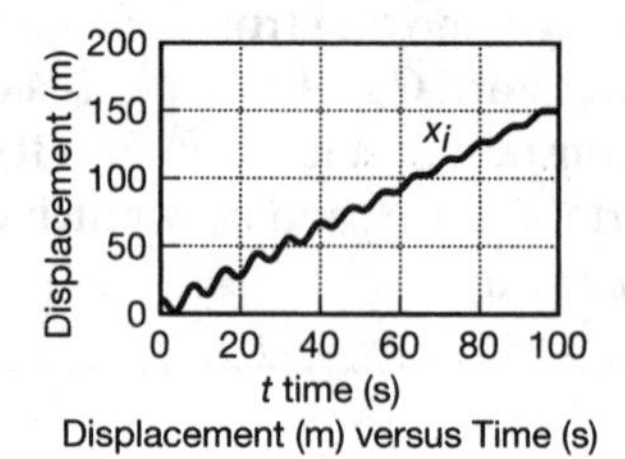

Displacement (m) versus Time (s)

$$v_w = -Ve^{-cx} \sin(\omega t),$$

where V m/s is the maximum wave velocity, c m^{-1} is the decay of the wave velocity, and ω s^{-1} is the frequency of the wave. Determine the absolute velocity of the surfer and the time required for her to paddle out 150 m from the beach, in terms of the quantities v, c, and ω.

This problem just requires the solution of a single first-order differential equation of motion. Following the procedures used to solve the previous sample problems, we prepare the following m-file with the equation of motion:

```
function xdot=otwosix(t,x);
v=6;vbw=1.5;c=1/100;w=pi/2;
xdot=-v*exp(-c*x)*sin(w*t)+vbw;
```

Then we use the following series of commands to call the previously stated function, numerically solve it via a Runge–Kutta method, and plot the results:

```
EDU>tspan=[0 100];
EDU>x0=10;
EDU>ode45('otwosix',tspan,x0)
```

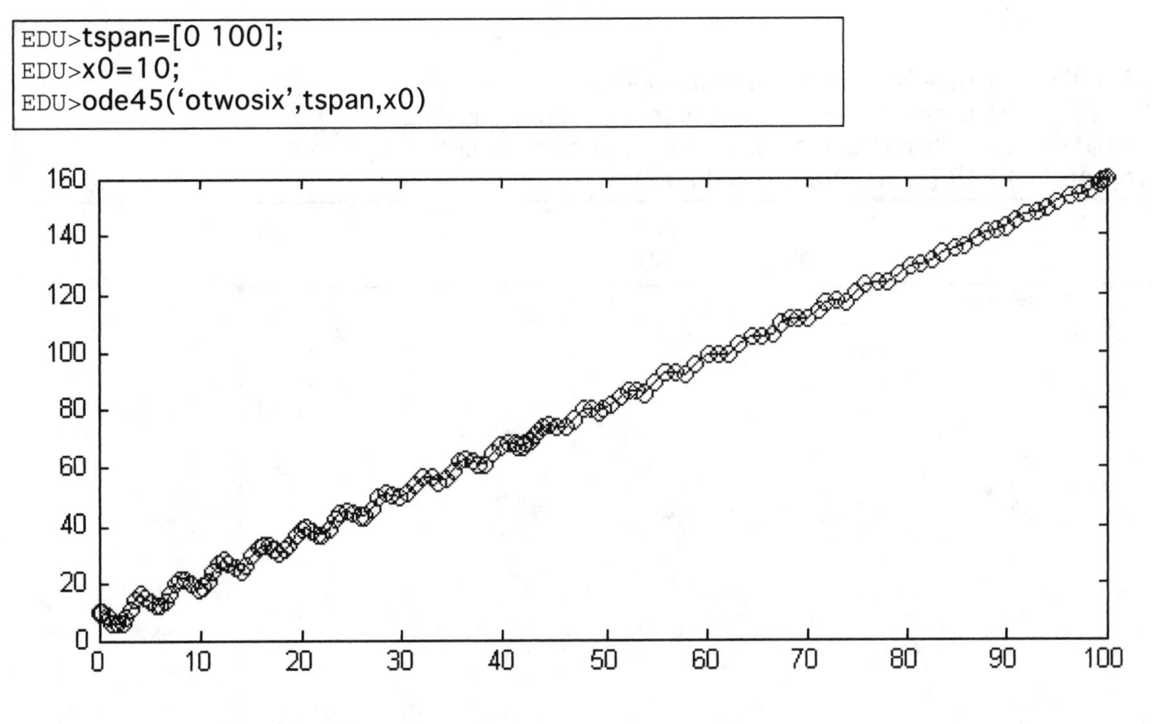

Sample Problem 1.34

A car starting from rest accelerates at a constant rate of 10 m/s^2, driving in a circle of radius 1000 m. Determine how far and how long the car will go before the normal acceleration component equals the tangential acceleration.

This sample problem can be solved by first making the computations symbolically, then manually converting the symbolic expressions into numerical ones, and finally plotting them:

```
EDU>syms t          % define t as symbolic
EDU>s=5*t^2;        % define s
EDU>g=s/1000;v=diff(s);     % define theta and the velocity
EDU>s               % display s for copying
```

```
s =
5*t^2
EDU>an                    % display an for copying
an =
1/10*t^2
EDU>t=(0:.5:10);          % redefine t as a numerical value
EDU>sn=5*t.^2;            % paste in s and make it numerical, changing ^ to .^
EDU>ann=1/10*t.^2;        % paste in an and make it numerical, changing ^ to .^
EDU>plot(sn,ann,'*'),title('an(t) versus s(t)')
```

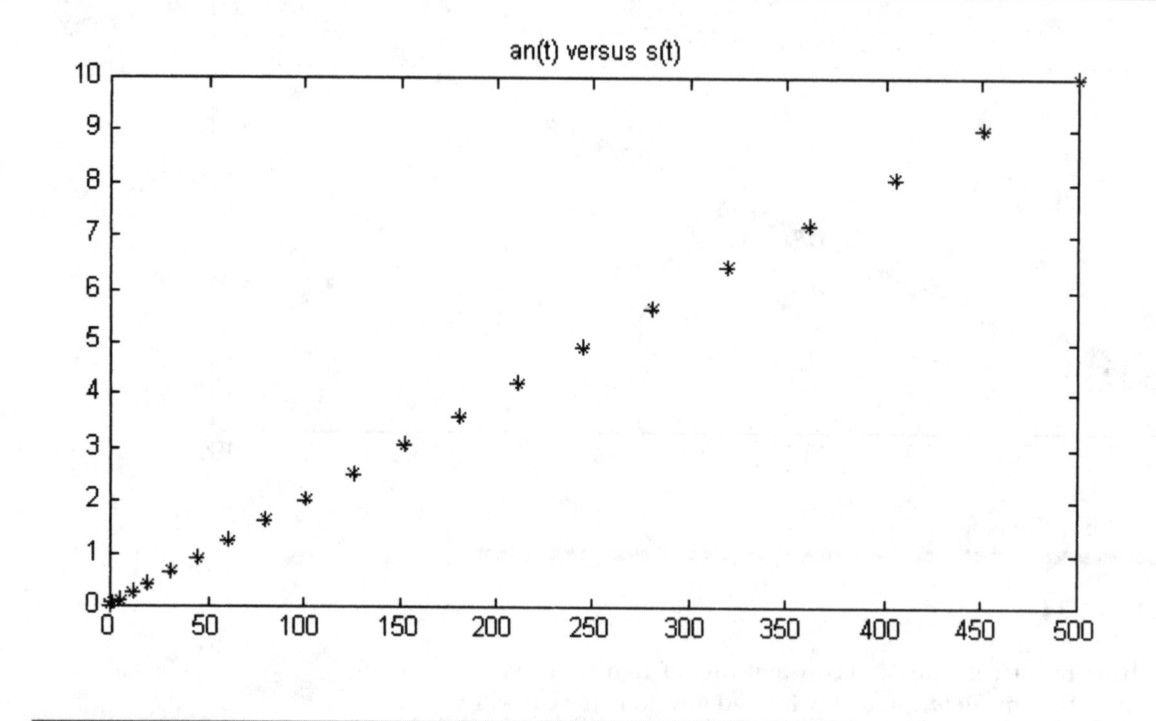

Sample Problem 1.35

Solve Sample Problem 1.34 for an automobile following a curved path given by

$$\theta(s) = \frac{s}{400}\sin\left(\frac{s}{200}\right).$$

This sample problem is solved symbolically by using the following commands:

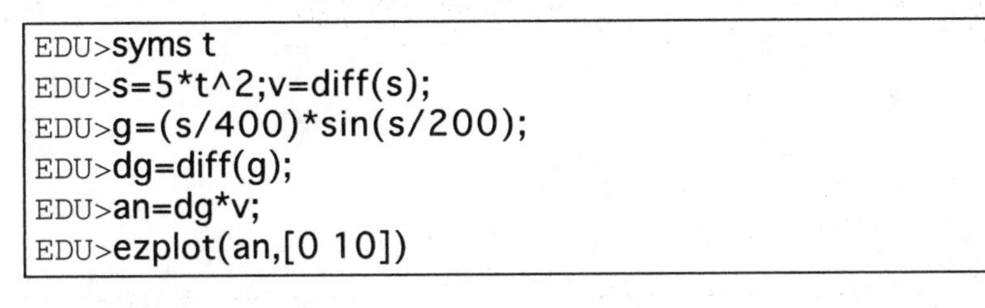

```
EDU>syms t
EDU>s=5*t^2;v=diff(s);
EDU>g=(s/400)*sin(s/200);
EDU>dg=diff(g);
EDU>an=dg*v;
EDU>ezplot(an,[0 10])
```

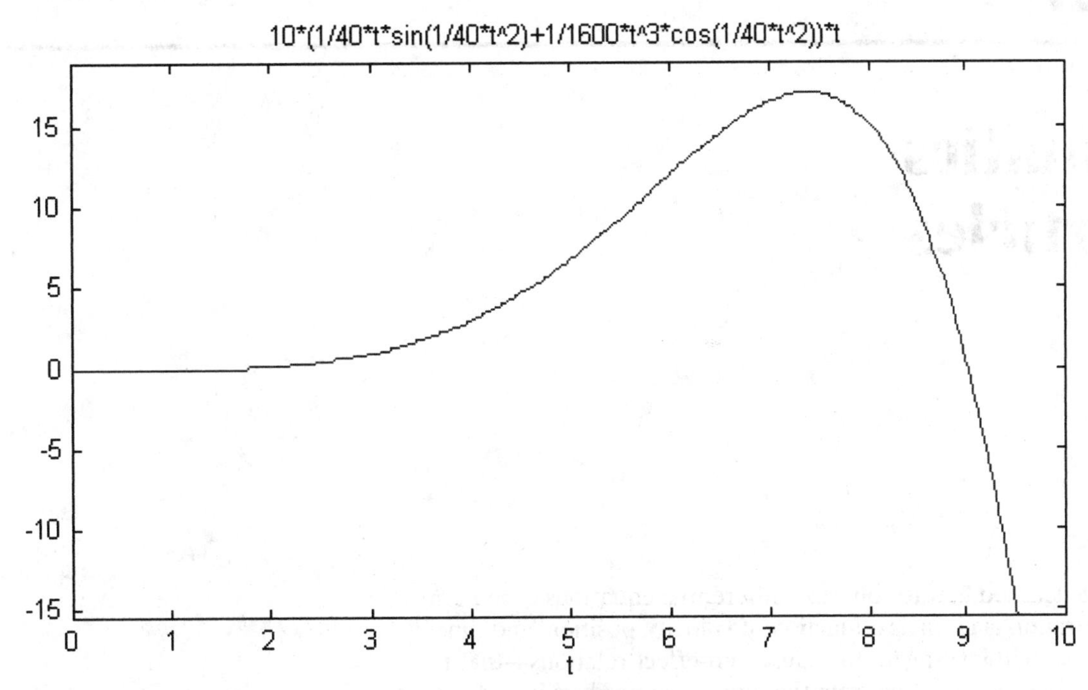

Next, the following command is entered to find out when the normal acceleration reaches 10 and to evaluate the function s at this time (recall the solve command works on the form an - 10 = 0):

```
EDU>solve(an-10)
ans =
        [ -5.620562279246527871449237986578O]
        [  5.620562279246527871449237986578O]
EDU>double(compose(s,5.620562279246527871449237986578O))
ans =
        157.9536
```

Note that the solution of the equation an = 10 has two solutions for time—one that is positive and one that is negative; we are interested in the positive value.

2

Kinematics of Particles

In Chapter 1, we learned how to solve the differential equations of motion when the acceleration is given as a function of velocity, position, and time. In this chapter, we will investigate the cause-and-effect relations—that is, how to determine the acceleration from the forces that produce it.

Sample Problem 2.1

The 3000-pound car shown in the accompanying figure is driving down a 30-degree incline when the driver locks up all wheels in a panic stop. If the coefficient of kinetic friction between the tires and the pavement is 0.7, how far will the car skid before coming to a stop if its initial speed was 45 mph?

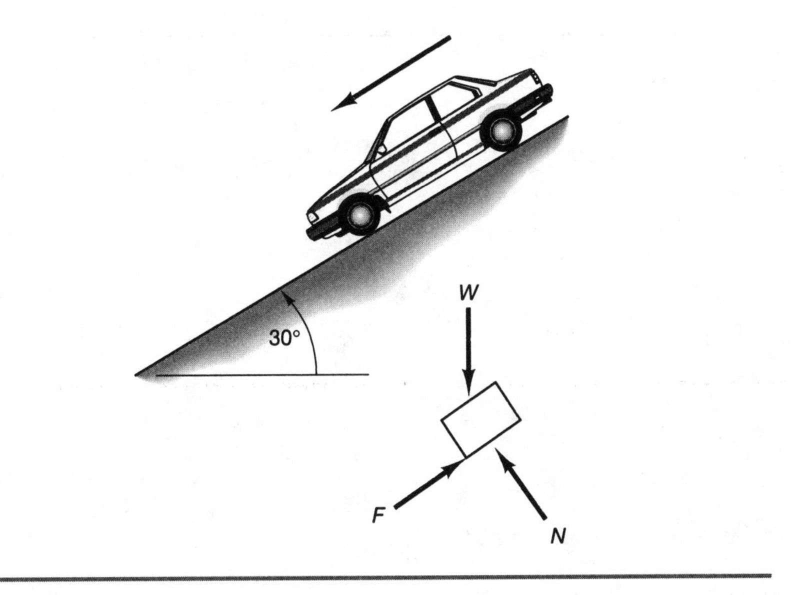

The acceleration and the stopping distance of the car on the incline are independent of the mass of the car. Therefore, the acceleration is dependent only on the incline angle and the coefficient of kinetic friction. First, consider using MATLAB to graphically represent a comparison between stopping distance and velocity for various types of surface. The governing equation is for the stopping distance is:

$$s(\theta, v_0, \mu_k) = \frac{v_0^2}{2g(\mu_k \cos\theta - \sin\theta)}$$

The following MATLAB code plots this function for various values of μ_k for a flat surface. Note that the code to produce the plot is fairly simple. The rest of the code illustrates numerous commands to produce a professional quality, well-labeled plot.

```
% Clear the workspace
clear all;
% Define system parameters
% Gravity
g = 32.2;
% Slope angle
theta = 0;
% Coefficients of friction
% Coefficients are in the following order
% Dry surface -- Wet Surface -- Snow surface -- Icy surface
mu_k = [0.7 0.5 0.3 0.1];
% Velocity values
```

```
v0 = [0:0.5:90];
% calculate the stopping distance

for i = 1:length (v0)
    for j = 1:length (mu_k)
        s(i,j) = v0(i)^2/(2*g*(mu_k(j)*cos(theta)-
sin(theta)));
     end;
end;

% Plot the results
figure;
plot(v0,s(:,1),'b-',v0,s(:,2),'r--',v0,s(:,3),'g-',
v0,s(:,4),'k--',...
   'LineWidth',2);
set(gcf,'Color','White');
set(gca,'FontName','times new roman', 'FontSize',20);
xlabel('Velocity (ft/sec)');
ylabel('Stopping Distance (ft)');
grid on;

text('Interpreter','latex',...
 'String','Ice',...
 'Position',[80 1200],...
 'FontSize',20)

text('Interpreter','latex',...
 'String','Snow',...
 'Position',[80 460],...
 'FontSize',20)

text('Interpreter','latex',...
 'String','Wet',...
 'Position',[87 285],...
 'FontSize',20)

text('Interpreter','latex',...
 'String','Dry',...
 'Position',[87 120],...
 'FontSize',20)
```

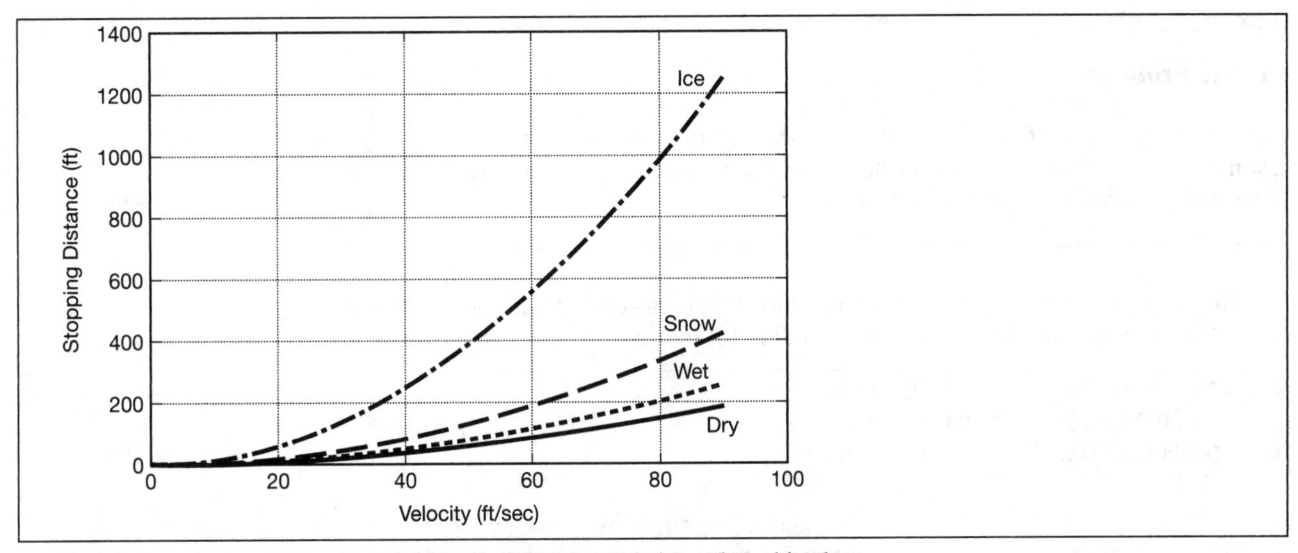

Recall that the concept of mesh plots in MATLAB is introduced in the supplement *Statics*. The "meshgrid" command must be used to prepare the x and y data for use in the "mesh" command, and the functional values must be computed using the prepared data (X and Y).

```
EDU>a=(0:0.1:pi/4);          % the range of values of theta
EDU>mu=(0.2:0.01:20/25);     % the range of values of the coefficient
EDU>[X,Y]=meshgrid(a,mu);    % prepares the previous values for plotting
EDU>Z=32.2*(sin(X)-Y.*cos(X));     % compute the acceleration
EDU>mesh(X,Y,Z)    % plot the mesh with Z "up"
```

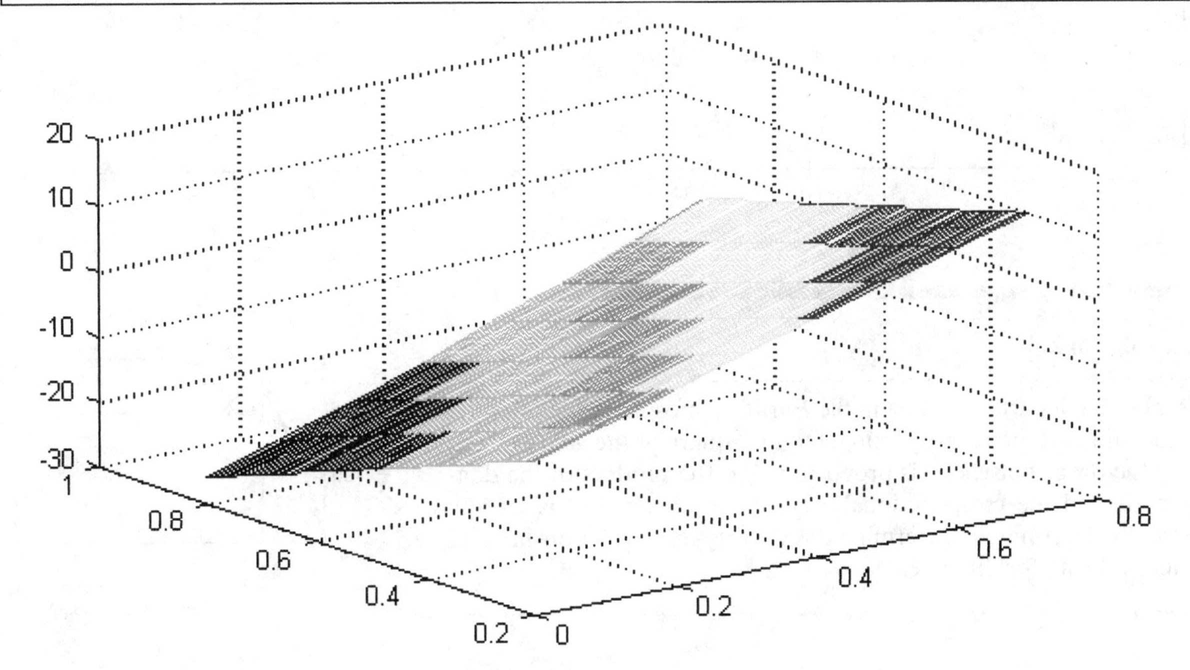

Sample Problem 2.3

Consider a particle of mass m falling in the Earth's atmosphere, and assume that air resistance is proportional to the velocity of the particle. Determine the velocity as a function of time.

$W = mg$

x

$F = cv$

This problem is a straightforward application of previously introduced commands to compute a function of time and to plot that function.

```
EDU>m=5;c=0.6;g=32.2;t=(0:.1:40);
EDU>v=(m*g/c)*(1-exp(-(c/m)*t));
EDU>plot(t,v),title('velocity versus time')
```

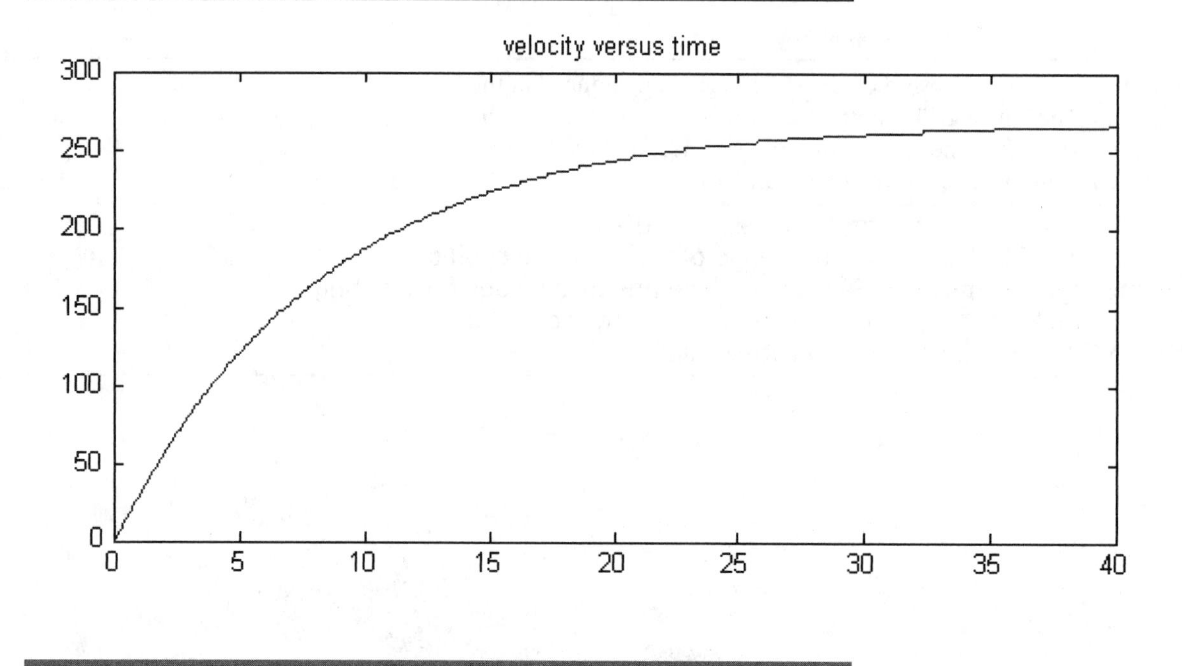

Sample Problem 2.4

A better model of a particle falling in the Earth's atmosphere includes the fact that the air resistance is proportional to the square of the velocity. The constant c, the drag coefficient, is proportional to the product of the density of the air and the cross-sectional area of the object, and is usually experimentally determined. Determine the velocity–time relationship and the terminal velocity for this case.

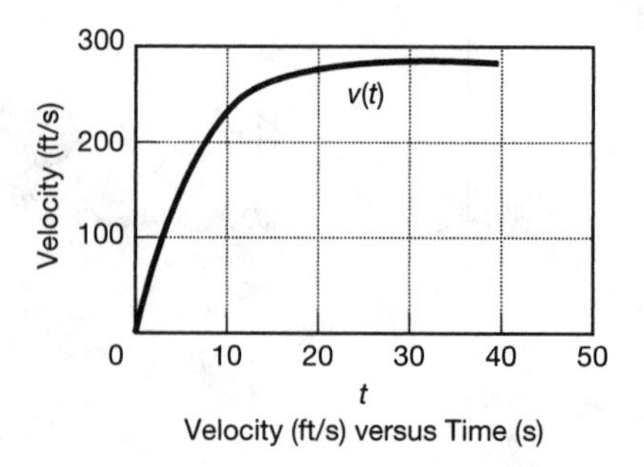

Velocity (ft/s) versus Time (s)

The solution to this sample problem is identical to that for the previous sample problem as far as the MATLAB code goes:

```
EDU>m=5;c=0.002;g=32.2; k=sqrt(c*g/m);     % enter the physical parameters
EDU>t=(0:0.5:40);     % define the range of time
EDU> v=(g/k)*(1-exp(-2*k*t))./(1+exp(-2*k*t));     % enter the expression for the velocity
EDU>plot(t,v,'*'),title('velocity versus time')     % plot v versus t
```

The previous commands produce the following plot, which indicates a terminal velocity of about 25 seconds:

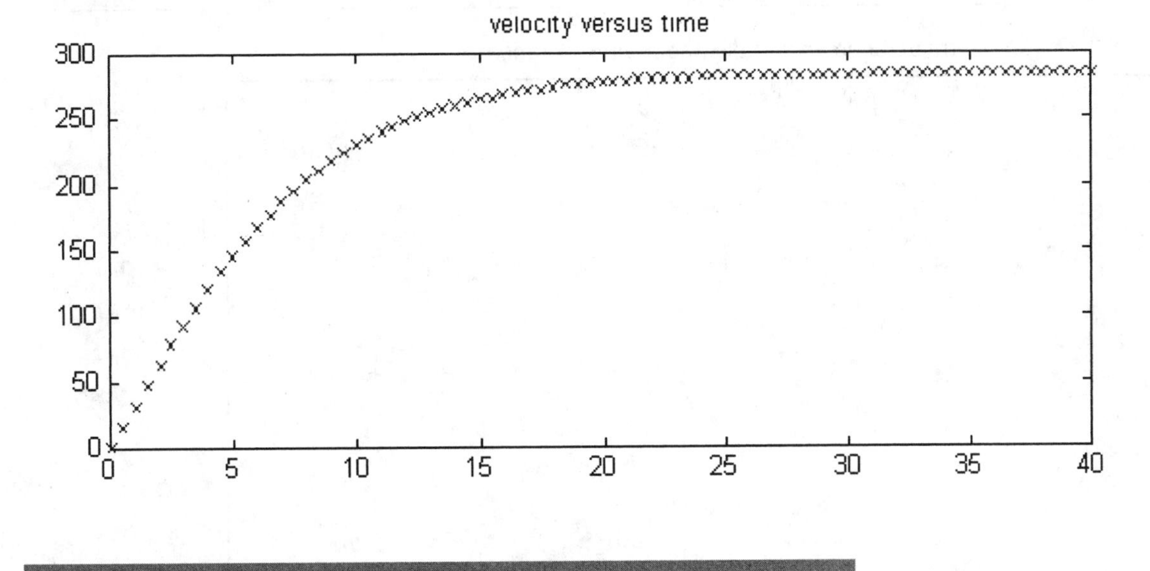

Sample Problem 2.5

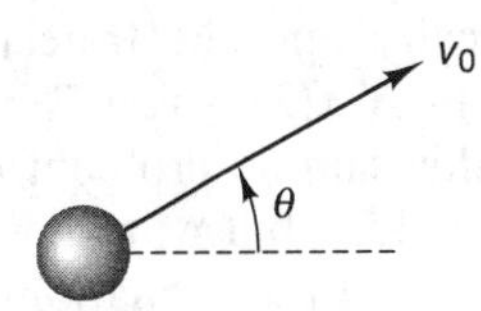

(a) Compare the distance that a ball can be thrown when air resistance is proportional to the velocity with the distance the ball can be thrown when air resistance is ignored. The ball has an initial velocity of v_0 at an angle θ with the horizontal.

(b) Write the equations of motion if the air resistance is proportional to the square of the velocity, and discuss the difficulties of solving the equations.

This sample computes a trajectory and plots it. However, in the following code, two different trajectories are plotted:

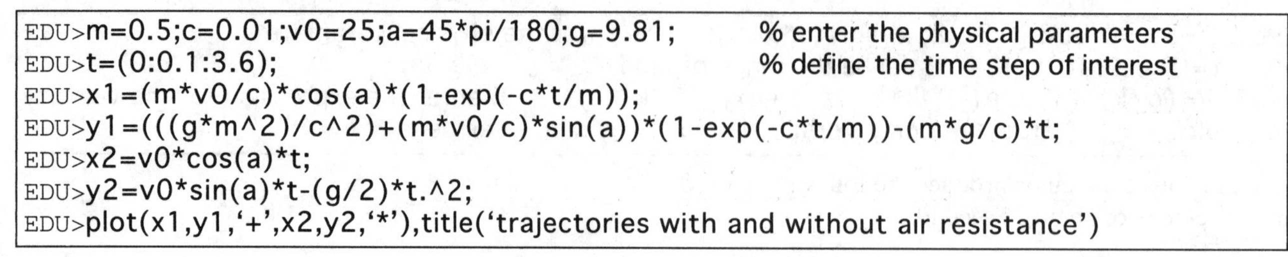

```
EDU>m=0.5;c=0.01;v0=25;a=45*pi/180;g=9.81;          % enter the physical parameters
EDU>t=(0:0.1:3.6);                                    % define the time step of interest
EDU>x1=(m*v0/c)*cos(a)*(1-exp(-c*t/m));
EDU>y1=(((g*m^2)/c^2)+(m*v0/c)*sin(a))*(1-exp(-c*t/m))-(m*g/c)*t;
EDU>x2=v0*cos(a)*t;
EDU>y2=v0*sin(a)*t-(g/2)*t.^2;
EDU>plot(x1,y1,'+',x2,y2,'*'),title('trajectories with and without air resistance')
```

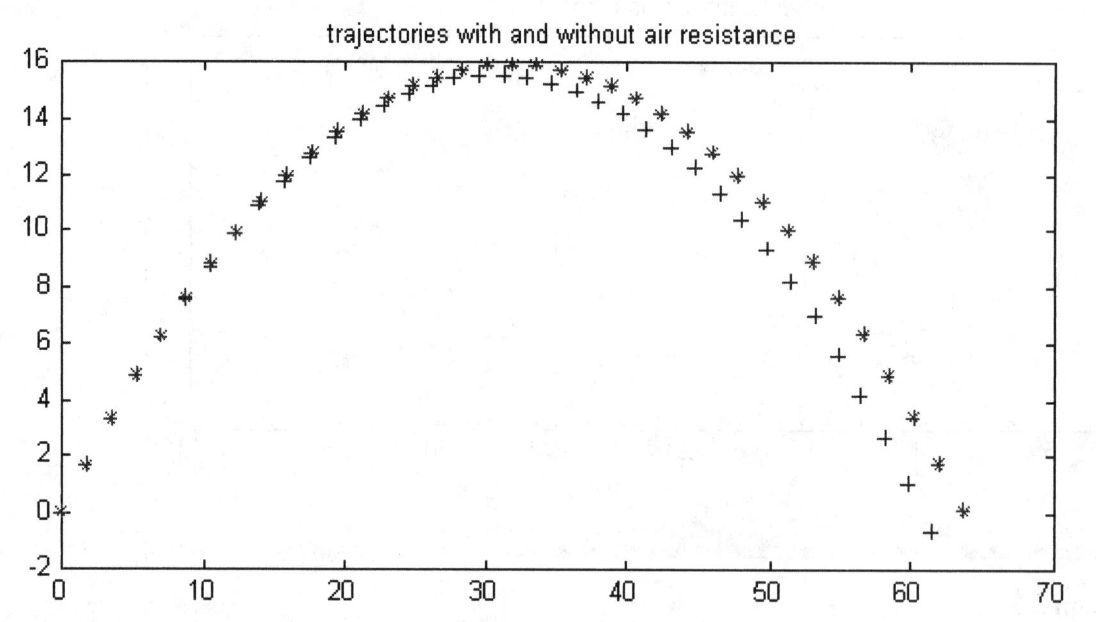

Next, the equations of motion for a nonlinear drag force are coded, and solved numerically, and the trajectories are plotted. The variables are renamed as $x(1) = x, x(2) = v_x, x(3) = y$, and $x(4) = v_y$. The MATLAB code used to solve this sample problem is exactly like the code used to solve Sample Problem 1.24, but with new differential equations. The following m-file is prepared with the equations of motion:

```
function xdot=twopt5(t,x);
g=32.2;c=0.0001;m=1/(2*g);     % enters the physical constants
xdot=[x(2);-(c/m)*x(2)*sqrt(x(2)^2+x(4)^2);x(4);-(c/m)*x(4)*sqrt(x(2)^2+
x(4)^2)-g];
```

Next, the following commands are used to call and solve the equations of motion and to plot the trajectory:

```
EDU>tspan=[0 2.5];
EDU>a=45*pi/180;V=60;
EDU>x0=[0;V*sin(a);0;V*cos(a)];
EDU>[t,x]=ode45('twopt5',tspan,x0);
EDU>plot(x(:,1),x(:,3))
EDU>plot(x(:,1),x(:,3)),title('x-y trajectory')
```

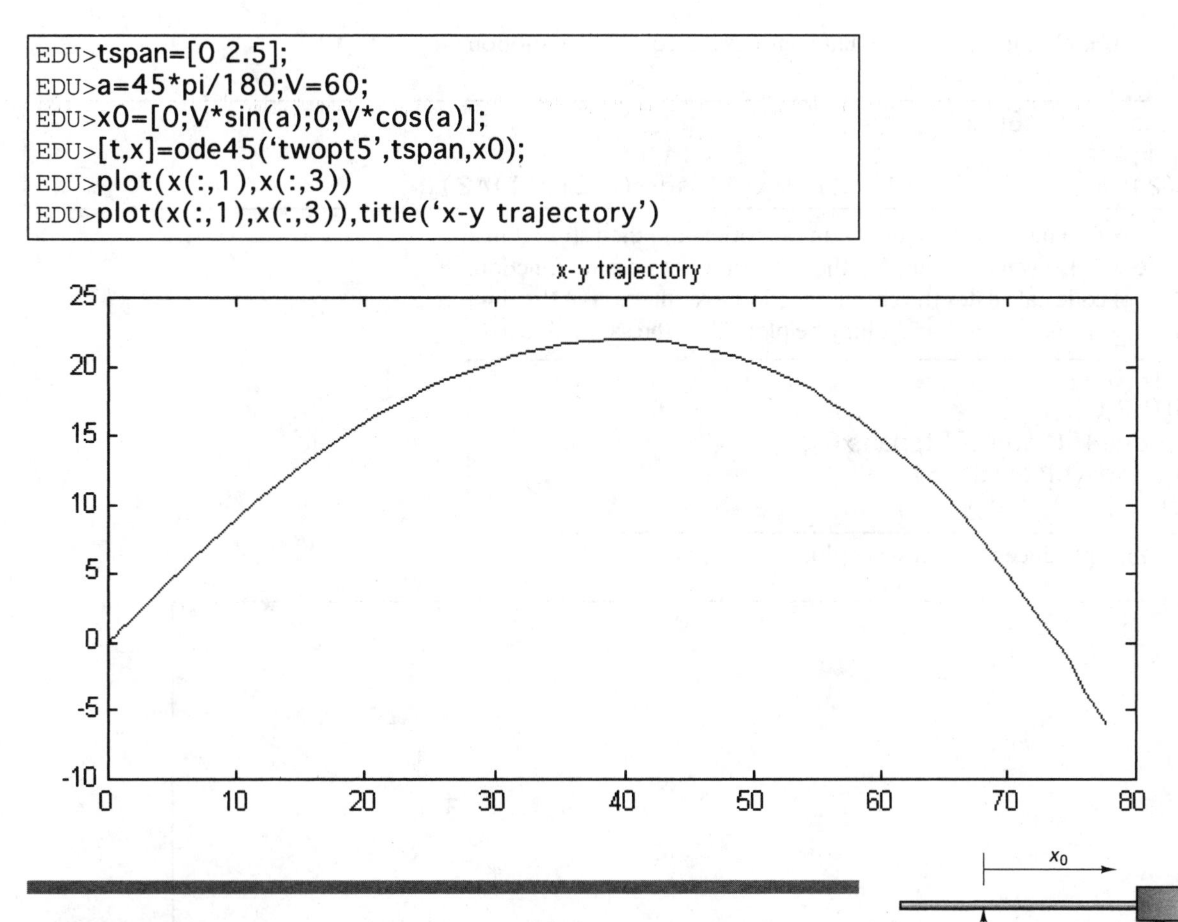

Sample Problem 2.6

A spring slider mechanism is used to produce oscillatory motion. If the unstretched length of the spring, which has a spring constant k, is l, and the slider of mass m is released from rest at the position shown in the accompanying diagram, write the equation of motion. Neglect the mass of the spring.

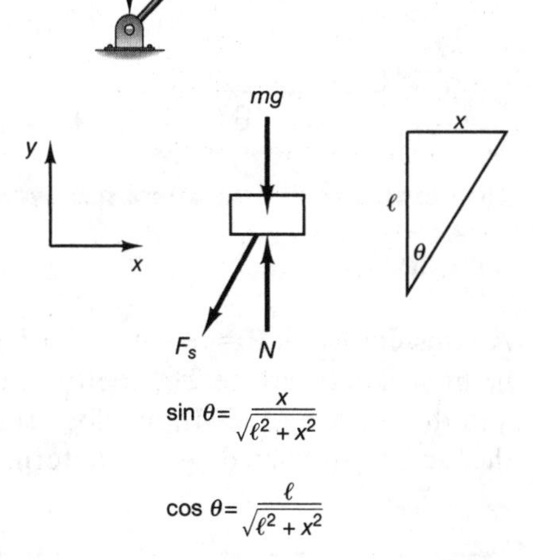

The solution to this sample problem involves solving a single nonlinear differential equation of motion, as done to solve Sample Problem 1.14. In this problem, however, we wish to plot the time response and to compare this response to a standard cosine response, so the code for Sample Problem 1.14

is modified accordingly. First, the m-file that contains the equation of motion is prepared:

```
function xdot=twopt6(t,x);
k=200;l=1;
xdot=[x(2);-k*(sqrt(l^2+x(1)^2)-l)*x(1)/sqrt(l^2+x(1)^2)];
```

The next set of commands solve the equation of motion and then store it in a matrix **x**. Since we also want to compare the solution to the cosine function, the fourth line of code calculates the cosine function as a function of the time t generated in the m-file so that y and x may be plotted on the same time axis.

```
EDU>tspan=[0 4];
EDU>x0=[0.2;0];
EDU>[t,x]=ode45('twopt6',tspan,x0);
EDU>y=0.2*cos(pi*t/1.875);
EDU>plot(t,x(:,1),'*',t,y,'+')
```

These commands produce the following plot:

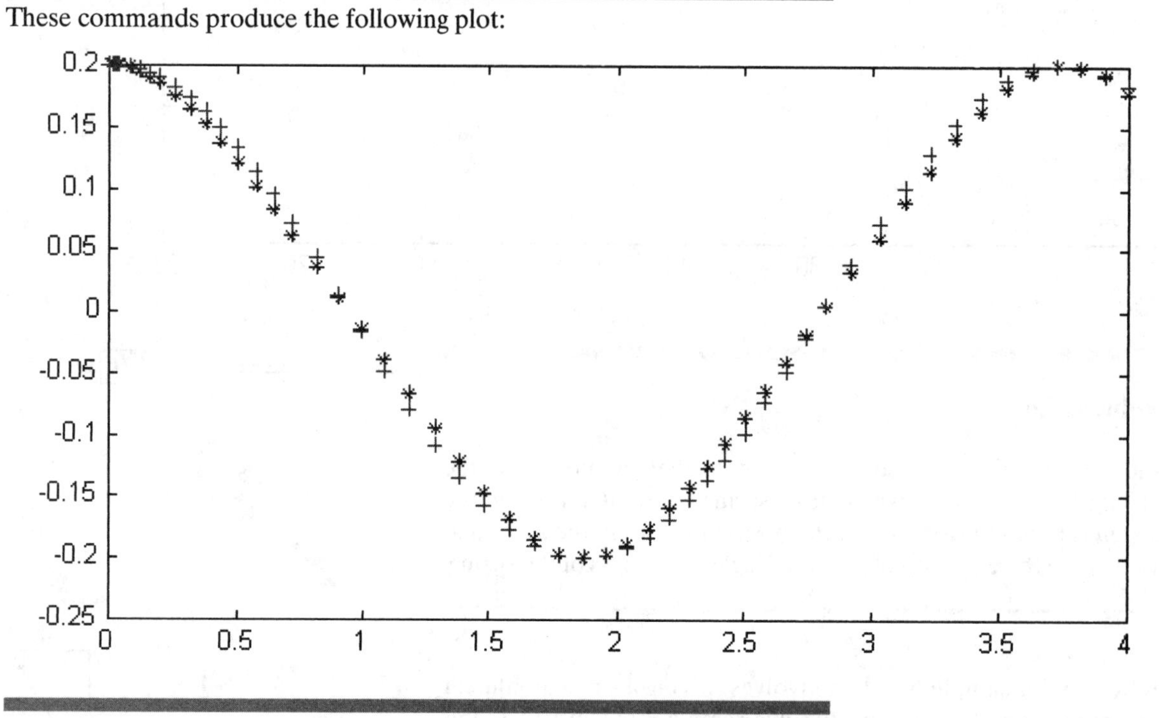

Sample Problem 2.7

A constant force $\mathbf{P} = -500\hat{\mathbf{i}} - 300\hat{\mathbf{j}}$ is applied to a 20-kg block to push it up an inclined surface. The coefficient of kinetic friction between the block and the surface is 0.3. If the block starts at rest from point D, as shown in the accompanying diagram, determine the position of the block after 2 seconds.

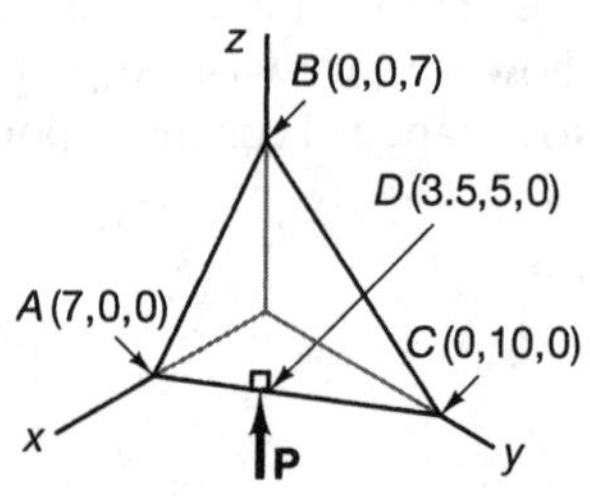

The vector calculations required for this sample problem can all be made in the command window. The time dependence can be added symbolically, as indicated in the following code:

```
EDU>AB=[-7;0;7];AC=[-7;10;0];
EDU>n=cross(AC,AB)/norm(cross(AC,AB))
n =
     0.6337
     0.4436
     0.6337
EDU>P=[-500;-300;0];W=[0;0;-20*9.81];
EDU>p=(P+W)/norm(P+W)
p =
    -0.8127
    -0.4876
    -0.3189
EDU>pn=dot(p,n)*n
pn =
    -0.5916
    -0.4141
    -0.5916
EDU>pt=p-pn
pt =
    -0.2212
    -0.0735
     0.2726
EDU>f=-pt/norm(pt)
f =
     0.6166
     0.2050
    -0.7601
EDU>r0=[3.5;5;0];
```

Next, the symbolic processor is used to define the vector **r** as a function of time, and then the vector is evaluated at $t = 2$ seconds:

```
EDU>syms t
EDU>r=-(0.5*2.44*t^2)*f+r0
r =
[    -8469496736669277/11258999068426240*t^2+7/2]
[ -4506132242517880 99/18014398509481984 00*t^2+5]
[      417637183710833947/450359962737049600*t^2]
EDU>r2=double(subs(r,t,2))
r2 =
     0.4910
     3.9994
     3.7094
```

Sample Problem 2.8

If the pulley system in the accompanying illustration is released from rest, determine the velocity of the 100-kg block C after it has moved 0.5 m. Neglect friction and the weight of the pulleys.

$$m_A = 25 \text{ kg} \quad m_B = 40 \text{ kg} \quad m_C = 100 \text{ kg}.$$

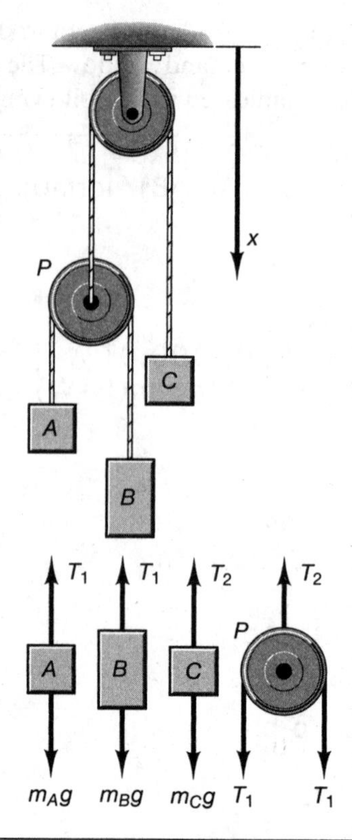

The solution to this sample problem uses the symbolic processor in MATLAB to solve a system of equations by using the matrix inverse. First, the variables are declared to be symbolic, and then the relevant matrix and vector are entered in terms of these symbolic variables. The solution is calculated using a simple matrix-inverse command, which produces the solution symbolically. Here, the solution vector is of the form $\mathbf{x} = [a_A \, a_B \, a_C \, T_1]^T$, as indicated in the *Dynamics* text.

```
EDU>syms ma mb mc g
EDU>A=[1 1 2 0;ma 0 0 1;0 mb 0 1;0 0 mc 2]
A =
[ 1,   1,   2, 0]
[ma,   0,   0, 1]
[ 0,  mb,   0, 1]
[ 0,   0,  mc, 2]
EDU>b=[0;ma*g;mb*g;mc*g];
EDU>x=inv(A)*b
x =
[   (mc+4*mb)/(mb*mc+ma*mc+4*ma*mb)*ma*g-3*mc/(mb*mc+ma*mc+4*ma*mb)*mb*g]
[  -3*mc/(mb*mc+ma*mc+4*ma*mb)*ma*g+(mc+4*ma)/(mb*mc+ma*mc+4*ma*mb)*mb*g]
[     -4*mb/(mb*mc+ma*mc+4*ma*mb)*ma*g+(mb+ma)/(mb*mc+ma*mc+4*ma*mb)*mc*g]
[                                   4*mb*mc/(mb*mc+ma*mc+4*ma*mb)*ma*g]
```

Next, the `subs` command of MATLAB is repeatedly used to produce a numerical solution for the specific set of values given in the text of the problem statement. The command `subs(x,ma,25)`, indicates that the value `ma` in the symbolic expression `x` is to be replaced with `25`.

```
EDU>x1= subs(x,ma,25);x2=subs(x1,mb,40);x3=subs(x2,mc,100);
     xn=double(subs(x3,g,9.81))
xn =
  -5.1386
   0.4671
   2.3357
 373.7143
```

The last expression uses the **double** command to convert the solution from a symbolic value into a numerical value.

Sample Problem 2.10

A 200-lb stuntman jumps from a 20-ft platform and lands on a mat to break his fall, as depicted in the accompanying illustration. The mat has a spring constant of 100 lb/ft and damping linearly proportional to the velocity of 40 lb s/ft. Note that the damping occurs only while the mat is being compressed. Write the equation of motion for the fall of the stuntman, and plot his motion with time. Determine the force exerted during contact with the mat.

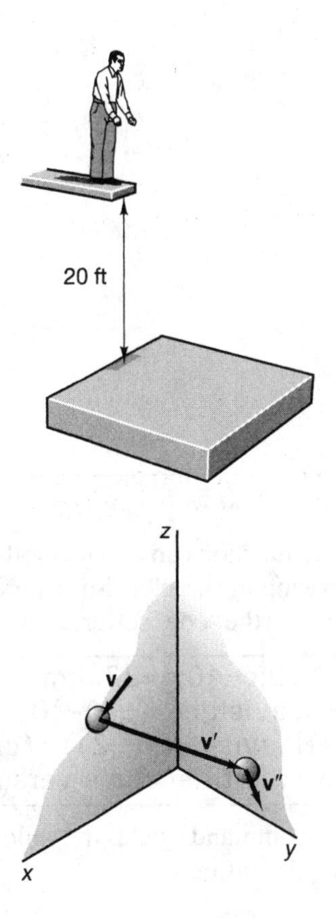

The solution to this problem involves the use of the Heaviside function and the solution of a differential equation. The Heaviside function is defined symbolically in MATLAB, but not numerically, per say. Hence, you must first define the Heaviside function by using another command, **stepfun**, which is defined in the Control System Toolbox. Thus, in addition to the m-file required to solve the equation of motion, we also need to establish the Heaviside function (unless you already have done so while studying the MATLAB supplement to *Statics*). The m-file for the Heaviside function is:

```
function y=Heaviside(t);
y=stepfun(t,0);
```

Next, the following m-file for the equations of motion is established:

```
function xdot=twopt10(t,x);
g=32.2;m=200/g;k=100;c=40;
xdot=[x(2);g-Heaviside(x(1)-20)*((k/m)*(x(1)-20)
     +Heaviside(x(2))*(c/m)*x(2))];
```

Running the following code in the command window solves the equation of motion and plots the solution:

```
EDU>tspan=[0 4];
EDU>x0=[0;0];
EDU>[t,x]=ode45('twopt10',tspan,x0);
EDU>plot(t,x(:,1)),title('response versus time')
```

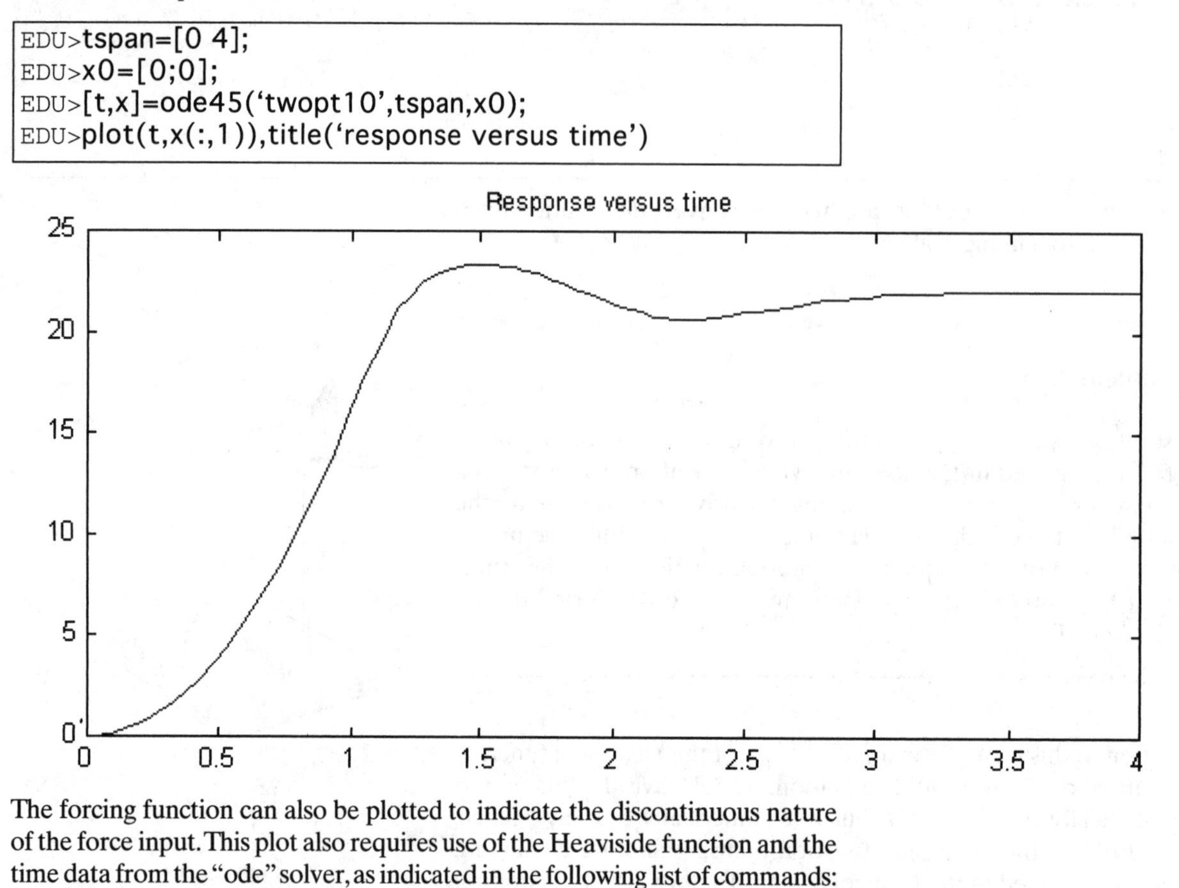

The forcing function can also be plotted to indicate the discontinuous nature of the force input. This plot also requires use of the Heaviside function and the time data from the "ode" solver, as indicated in the following list of commands:

```
EDU>k=100;c=40;g=32.3;m=200/g;
EDU>f=Heaviside(x(:,1)-20).*((k/m).*(x(:,1)-20)
      +Heaviside(x(:,2)).*(c/m).*x(:,2));
EDU>plot(t,f),title('force versus time')
```

This set of commands yields the following plot of the forcing function, indicating the discontinuity:

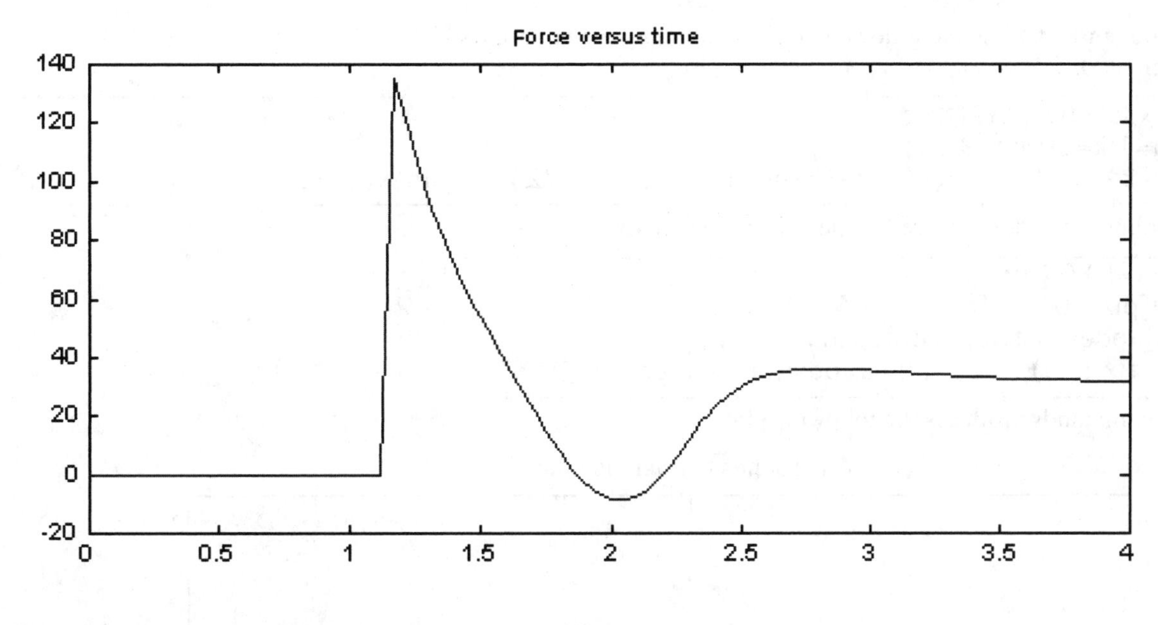

Sample Problem 2.12

Determine the differential equation of motion for the spring–slider mechanism shown in the accompanying diagram. Neglect friction between the slider and the curved bar, and neglect the mass of the spring. The slider of mass m is released from rest when $\theta = 60°$, and the unstretched length of the spring with spring constant k is R.

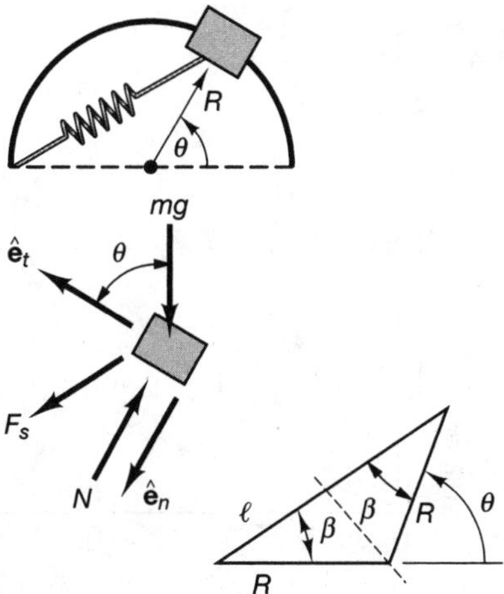

The answer to this problem is a straightforward solution of a single scalar differential equation, and it follows the steps of the solution to

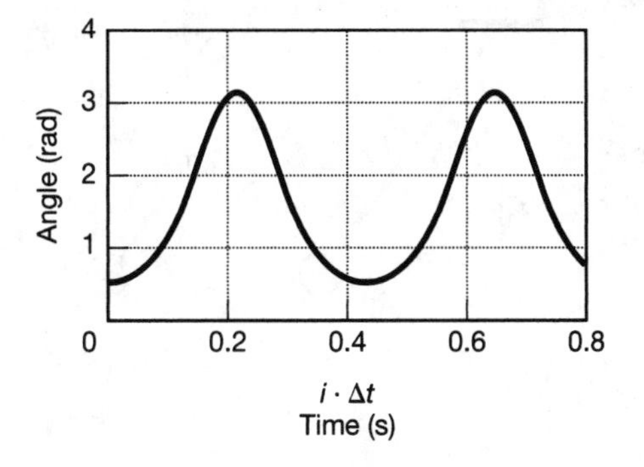

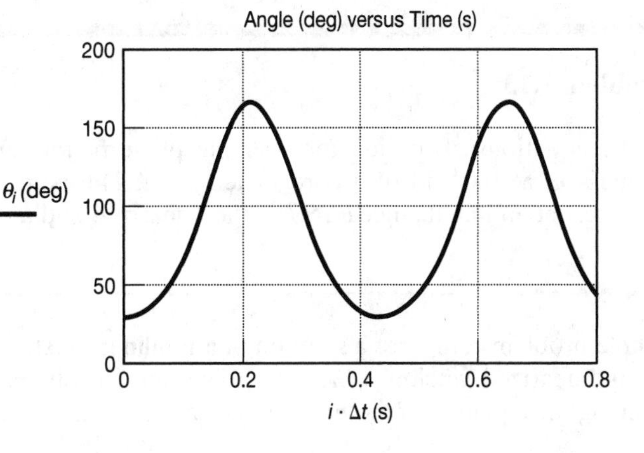

Sample Problem 1.14. First, the following m-file containing the equation of motion is created:

```
function xdot=twopt12(t,x);
R=0.2;m=1;k=600;g=9.81;
xdot=[x(2);-(g/R)*sin(x(1))+(k/m)*(2*cos(x(1)/2)-1)*sin(x(1)/2)];
```

Next, the following code is entered in the command window:

```
EDU>tspan=[0 0.8];
EDU>x0=[pi/6 0];
EDU>[t,x]=ode45('twopt12',tspan,x0);
EDU>plot(t,x(:,1)),title('angular position versus time')
```

This set of commands produces the following plot:

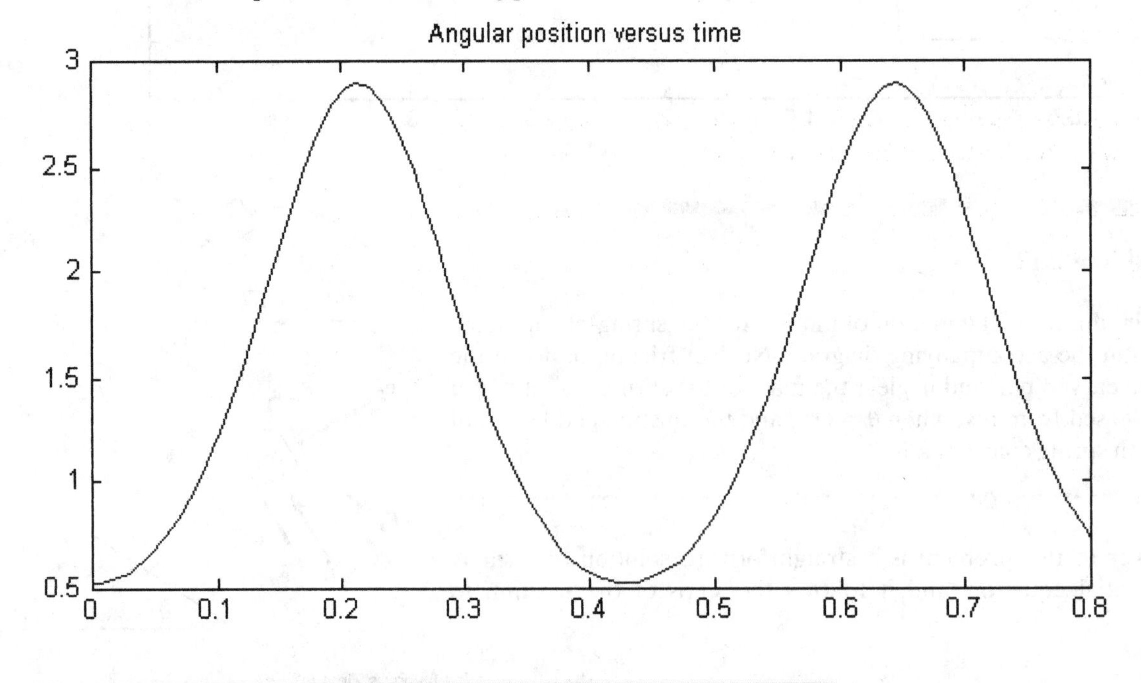

Sample Problem 2.13

Determine the equations of motion for a simple plane pendulum composed of a mass m at the end of a cord of length L. The pendulum is released from rest at an initial angle θ_0. Solve the linearized and nonlinear equations.

This sample problem compares a solution of a nonlinear system to the solution of its linearized version, which is known analytically. First, the m-file containing the equation of motion is entered:

```
function xdot=twopt13(t,x);
L=2;g=9.81;
xdot=[x(2);-(g/L)*sin(x(1))];
```

Next, this m-file is called to solve the nonlinear equation of motion. Then, after the time interval is established, the fifth line of code in the command window computes the analytical solution to the linearized equation. Finally, the linear and nonlinear solutions are plotted on the same graph to illustrate the differences between them.

```
EDU>tspan=[0 6];
EDU>x0=[1.5;0];
EDU>[t,x]=ode45('twopt13',tspan,x0);
EDU>g=9.81;L=2;
EDU>lin=1.5*cos(t*sqrt(g/L));
EDU>plot(t,x(:,1),'+',t,lin,'*')
```

These commands produce the following plot, in which you should note the difference between the linear and nonlinear response:

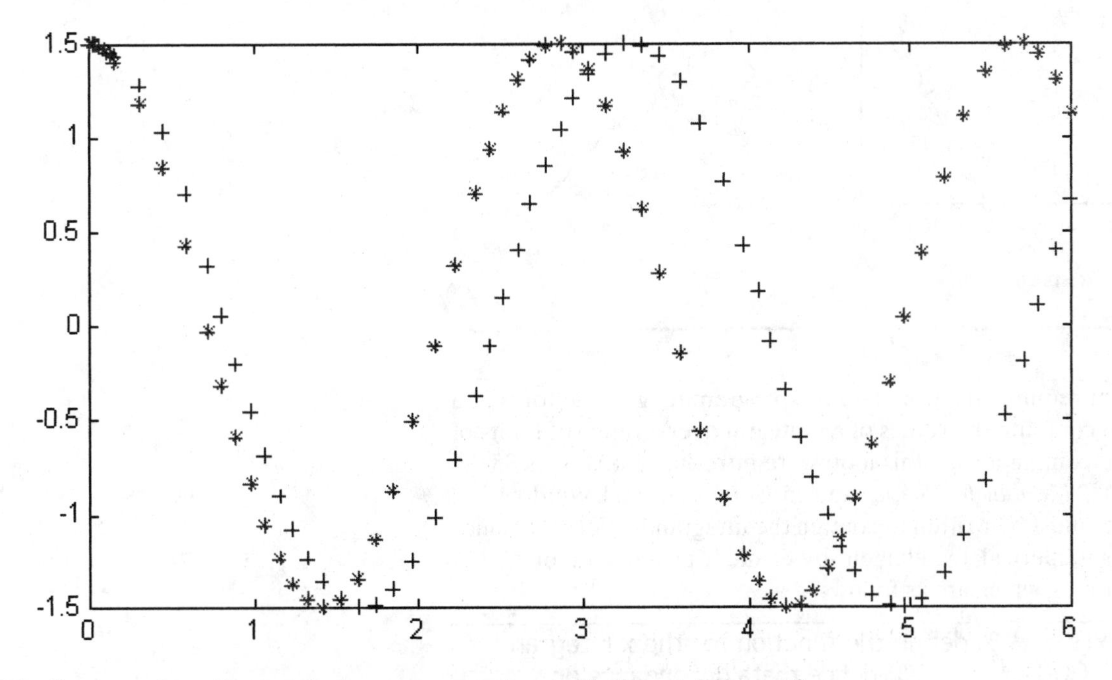

The linear and nonlinear solutions appear to be oscillating at different frequencies. Note that as the value of the initial displacement is reduced, the difference between the solutions becomes negligible.

Sample Problem 2.14

A driver wants to test-drive a vehicle in a tightening spiral path (see accompanying figure) at a constant speed on the salt flats of Utah. If the coefficient of static friction between the tires and the ground is 0.5 and the test driver is driving at 60 ft/s, determine the point at which the car will slide out of the curve if the curve is given by

$$\theta(s) = 1 + \left(\frac{s}{1000}\right)^2.$$

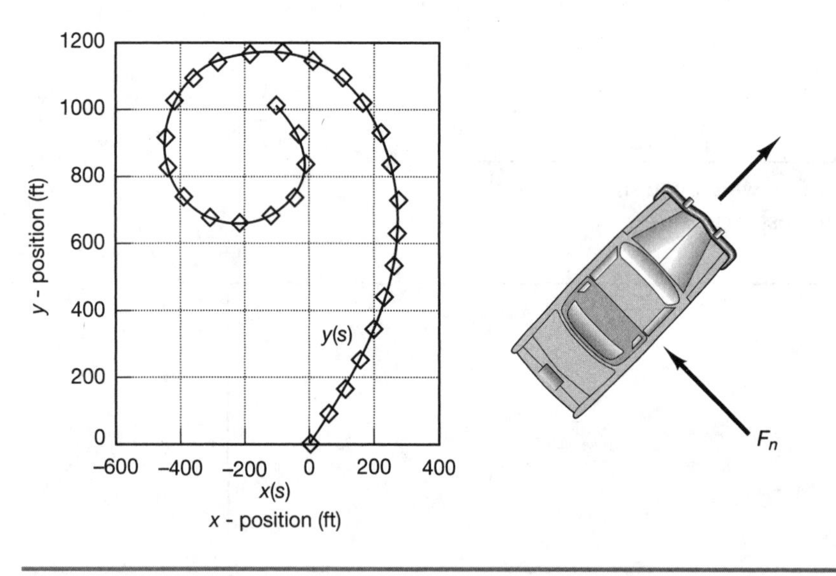

This problem requires the use of a little programming in the form of a loop in order to compute the values of an integral over a range of limits of integration. The commands for this loop were introduced on pages 85–86 of the *MATLAB Manual for Statics* and in Computational Window 5.2. First, two m-files must be written to contain the integrands. Note that since the variable **s** is numerical, the element-by-element power symbol "dot^" is used in the m-files, which are as follows:

```
function x=fx(s);    % define the function for the x integrand
theta=1+(s/1000).^2;       % define theta dependance on s
x=cos(theta);              % define the integrand
```

```
function y=fy(s);    % define the function for the y integrand
theta=1+(s/1000).^2;       % define theta dependance on s
y=sin(theta);              % define the integrand
```

Next, a third m-file is written to perform the integrations inside a loop and to store the integrations into two vectors, *xs* and *ys*. Remember that the notation "s(i)" refers to the *i*th element of the vector *s*. The m-file is named twopt14.

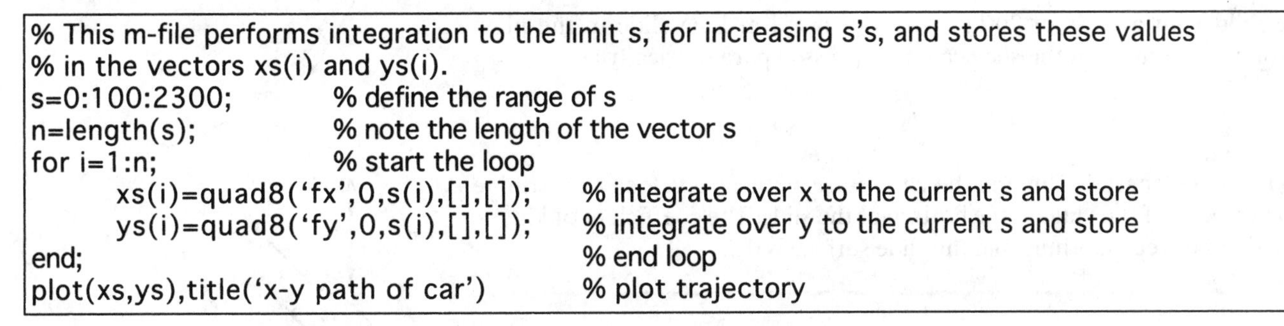

```
% This m-file performs integration to the limit s, for increasing s's, and stores these values
% in the vectors xs(i) and ys(i).
s=0:100:2300;              % define the range of s
n=length(s);               % note the length of the vector s
for i=1:n;                 % start the loop
      xs(i)=quad8('fx',0,s(i),[],[]);      % integrate over x to the current s and store
      ys(i)=quad8('fy',0,s(i),[],[]);      % integrate over y to the current s and store
end;                                       % end loop
plot(xs,ys),title('x-y path of car')       % plot trajectory
```

Typing twopt14 after the prompt produces the following plot of the path of the car:

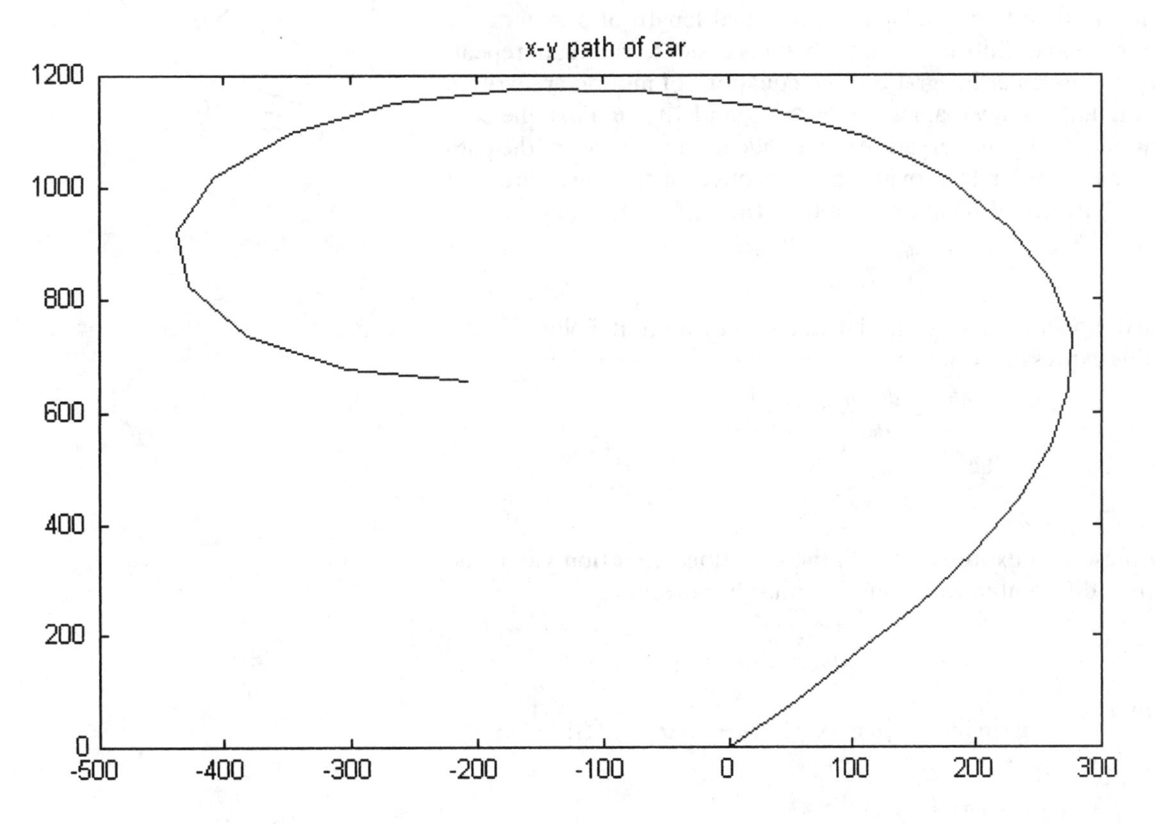

Sample Problem 2.15

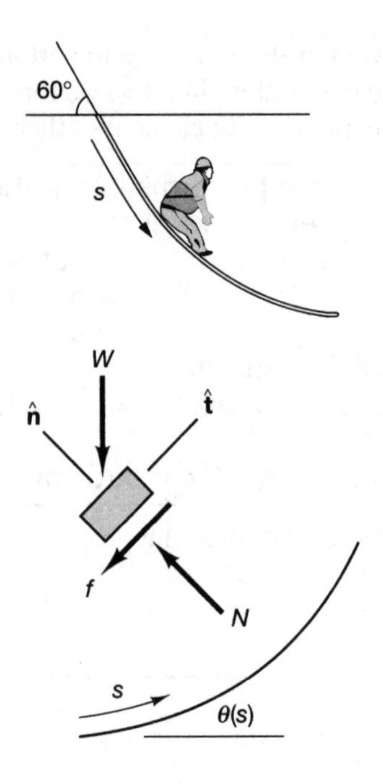

A child with a mass of 20 kg slides down a 3-m long slide in the playground. The curve of the slide can be expressed parametrically as

$$\theta(s) = -60°\left[1 - \left(\frac{s}{3}\right)^2\right].$$

If the slide length is 1 m and the child starts with zero velocity, determine the velocity of the child at the bottom of the slide. The coefficient of kinetic friction between clothing and the slide surface is 0.2.

In this problem, the equation of motion is nonlinear, and we would like to repeatedly solve the equations of motion over successively larger times until we find the time that corresponds to a final length of 3 meters. We first put the commands into an m-file so that it is easier to run them repeatedly, and we use an Euler integration. The equations of motion are written in terms of the following variables: `v` = v, `s` = s, and `th` = θ. First, the constraint equation relating the curvature of the slide to the distance of the path is differentiated in order to provide the derivative of the curvature that appears in the differential equation of motion. This differentiatian yields

$$\frac{d\theta}{ds} = \frac{2\pi}{9}s.$$

The derivative appears as $d\theta/ds$ in the equation of motion. Following the chain rule, this expression can be written as

$$\frac{d\theta}{ds} = \frac{d\theta}{dt}\frac{dt}{ds} = \frac{d\theta}{dt}\frac{1}{v}$$

so that

$$\frac{d\theta}{dt} = \frac{2\pi}{9}sv.$$

Combining these two expressions with the equation of motion yields the following three differential equations that must be solved:

$$\frac{ds(t)}{dt} = v(t)$$

$$\frac{dv(t)}{dt} = -g\sin(\theta(t)) - \mu\left[g\cos(\theta(t)) + \frac{2}{9}\pi^2(t)s(t)\right]$$

$$\frac{d\theta(t)}{dt} = \frac{2\pi}{9}s(t)v(t).$$

The m-file shown in the next box of code solves the equations of motion and the parametric equation simultaneously by using an Euler integration and the `for` command to perform the integration. This "for" command is run for increasing values of `N`, until the value of `s` reaches `3` meters.

```
% Sample Problem 2.15 using 3 states and an Euler integration
mu=.2;g=9.81;N=925;dt=.001;     % enter the physical constants and time interval
s(1)=0;v(1)=0;th(1)=-pi/3;      % enter the initial conditions
for n=1:N
th(n+1)=th(n)+(2*pi/9)*dt*(s(n).^2-1);
s(n+1)=s(n)+v(n)*dt;
v(n+1)=v(n)-g*sin(th(n))*dt-mu*(g*cos(th(n))+v(n).^2.*s(n)*(2*pi/9))*dt;
end
plot(s,v),title('velocity vs length along slide')
```

This m-file is run by typing the file's name after the prompt. Note that it is important to set all of the variables to zero each time the m-file is run; otherwise, the values of the variables will start iterating from the last computed values, forming a larger vector.

```
EDU>s=0;v=0;th=0;
EDU>twopt15
```

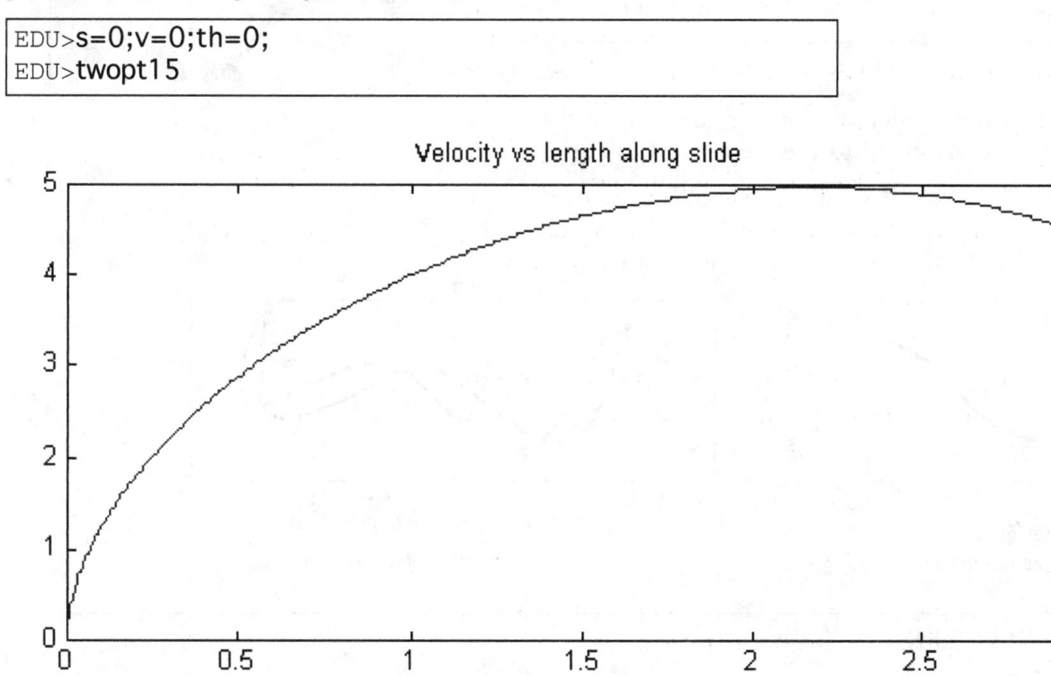

The values of s and v at the point where s = 3 m are determined by substitution of the value of N (an integer) into the vectors s and v to retrieve the numerical value of the last element of each vector, corresponding to the values of the vectors at the final time. The following commands accomplish this task:

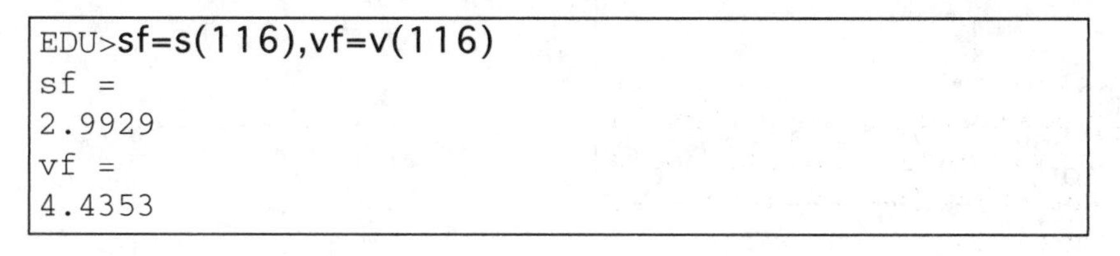

```
EDU>sf=s(116),vf=v(116)
sf =
2.9929
vf =
4.4353
```

Thus, the child is traveling 4.4353 m/s at the bottom of the slide. If friction is ignored, the equation of motion has an analytical solution that produces a final velocity of 6.03 m/s. The final time is

$$\mathrm{N} \times \mathrm{dt} = 925 \times 0.001 = 0.925 \text{ s.}$$

Sample Problem 2.17

A 2-kg mass is attached to a spring and released from rest at an angle $\theta = 30°$ and with the spring unstretched. If the spring's unstretched length is 300 mm and the spring constant is 1600 N/m, determine the motion of the mass.

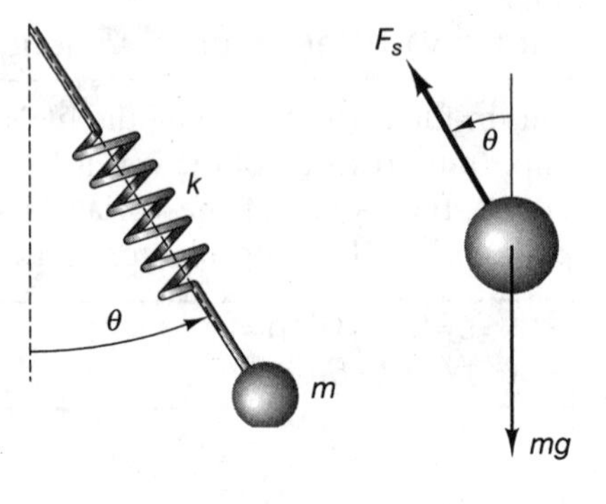

The solution to this is a straightforward application of the ode45 command, and its steps correspond to the steps used to solve Sample Problem 1.14. The m-file containing the equation of motion is:

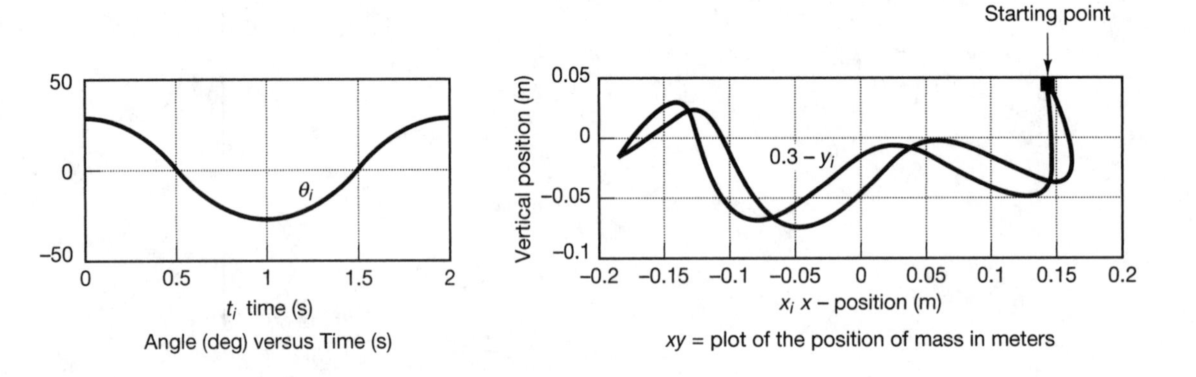


```
function xdot=twopt17(t,x);
m=2;k=1600;g=9.81;
xdot=[x(2);x(1)*x(4)^2+g*cos(x(3))-(k/m)*(x(1)-.3);x(4);
      (-1/x(1))*(g*sin(x(3))+2*x(2)*x(4))];
```

The following commands produce the desired plot of the motion:

```
EDU>tspan=[0 1.2];
EDU>x0=[0.3;0;pi/6;0];
EDU>[t,x]=ode45('twopt17',tspan,x0);
EDU>xi=x(:,1).*sin(x(:,3));yi=-x(:,1).*cos(x(:,3));
EDU>plot(xi,yi),title('x-y trajectory')
```

The second-to-last line in the previous set of commands converts the output into rectangular coordinates so that the plot appears as follows:

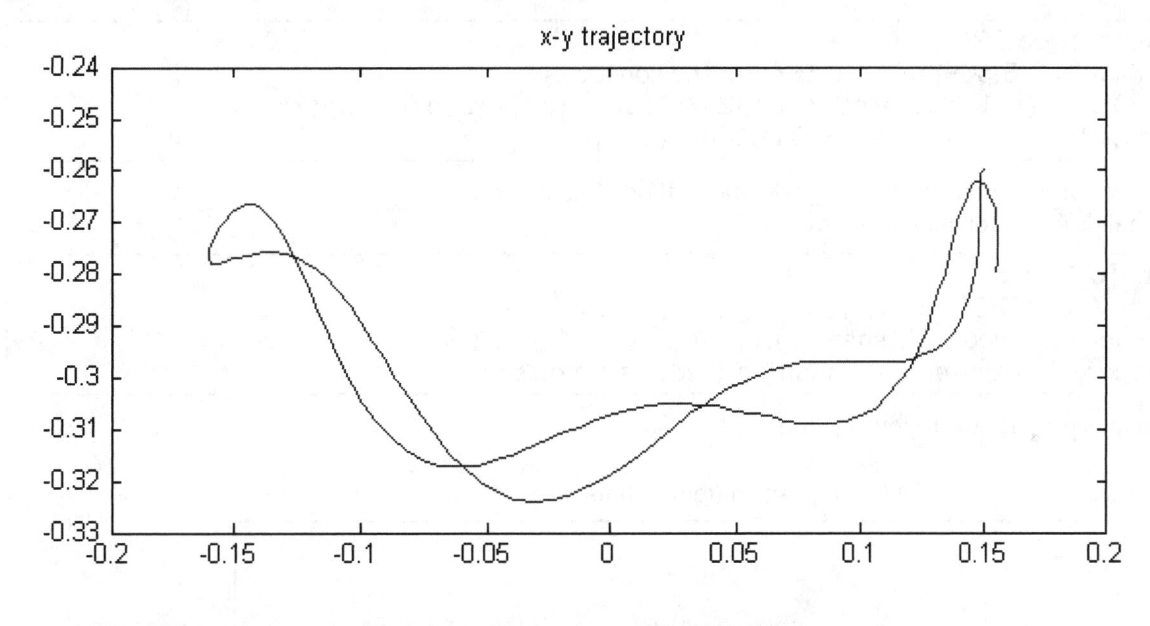

Sample Problem 2.22

A spherical pendulum is set into motion when $\phi = 30°$, with an initial velocity in the θ direction of 0.5 m/s. If the pendulum has a length of 2 m and a mass of 3 kg, determine its motion for the first 5 seconds. (See accompanying figure.)

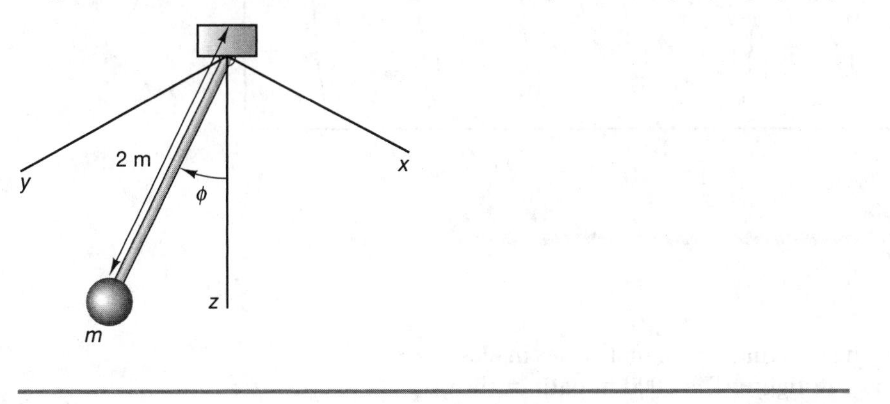

The solution to this problem involves running the "ode45" command in the manner that it was run in the solution to the previous sample problem.

The m-file for the equations of motion in first-order form, where $x(1) = \phi$, $x(2) = d\phi/dt, x(3) = \theta$, and $x(4) = d\theta/dt$, is as follows:

```
function xdot=twopt22(t,x);
R=2;g=9.81;v0=0.5;p0=pi/6;      % enter the constants
xdot=[x(2);(cos(x(1))*(sin(p0)*v0)^2)/(R*sin(x(1)))^2(g/R)*sin(x(1));
      x(4);-2*x(2)*x(4)*cos(x(3))/sin(x(1))];
```

To solve this set of equations using Runge–Kutta methods, enter the following commands in the command window:

```
EDU>tspan=[0 14];
EDU>x0=[pi/6;0;0;.25];
EDU>[t,x]=ode45('twopt22',tspan,x0);
EDU>plot(t,x(:,1)),title('angular position (rad) vs time (sec)')
```

These commands produce the following plot:

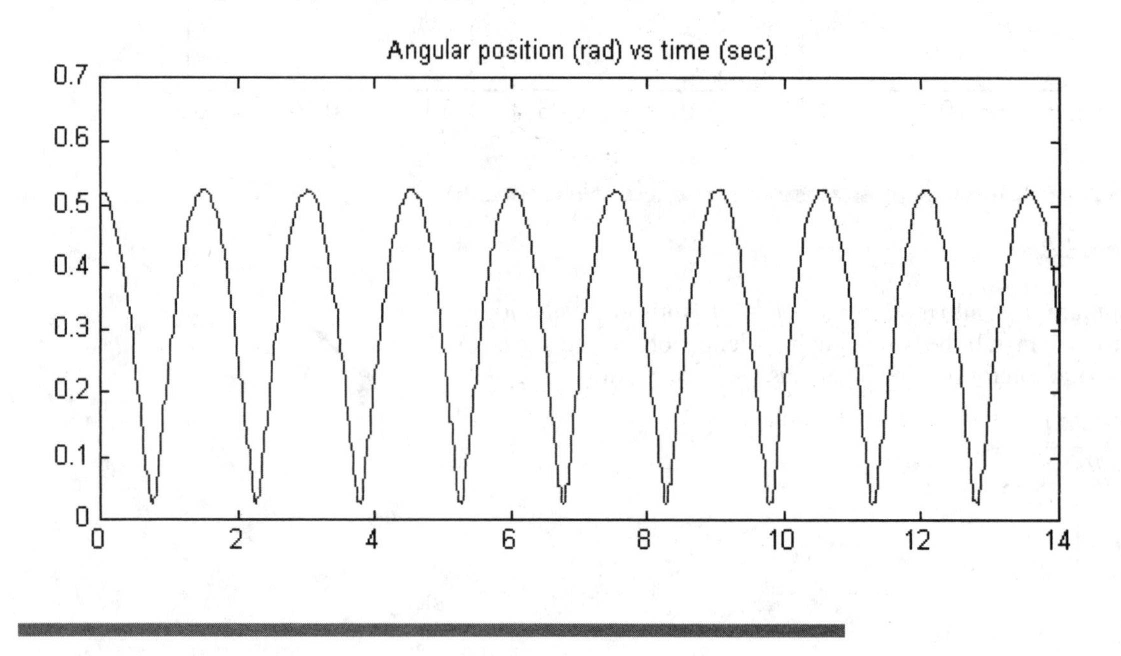

Sample Problem 2.23

A 1000-kg roller-coaster car moves along a circular track that varies in elevation in a sinusoidal manner as shown in figure. The 2000-m path of the roller coaster is given by

$$\theta(s) = \frac{\pi s}{1000} \qquad \beta(s) = \frac{\pi}{2}\sin\left(\frac{\pi s}{500}\right) \text{ rad.}$$

Determine the normal force exerted on the riders if the roller coaster moves at a constant velocity of 15 m/s.

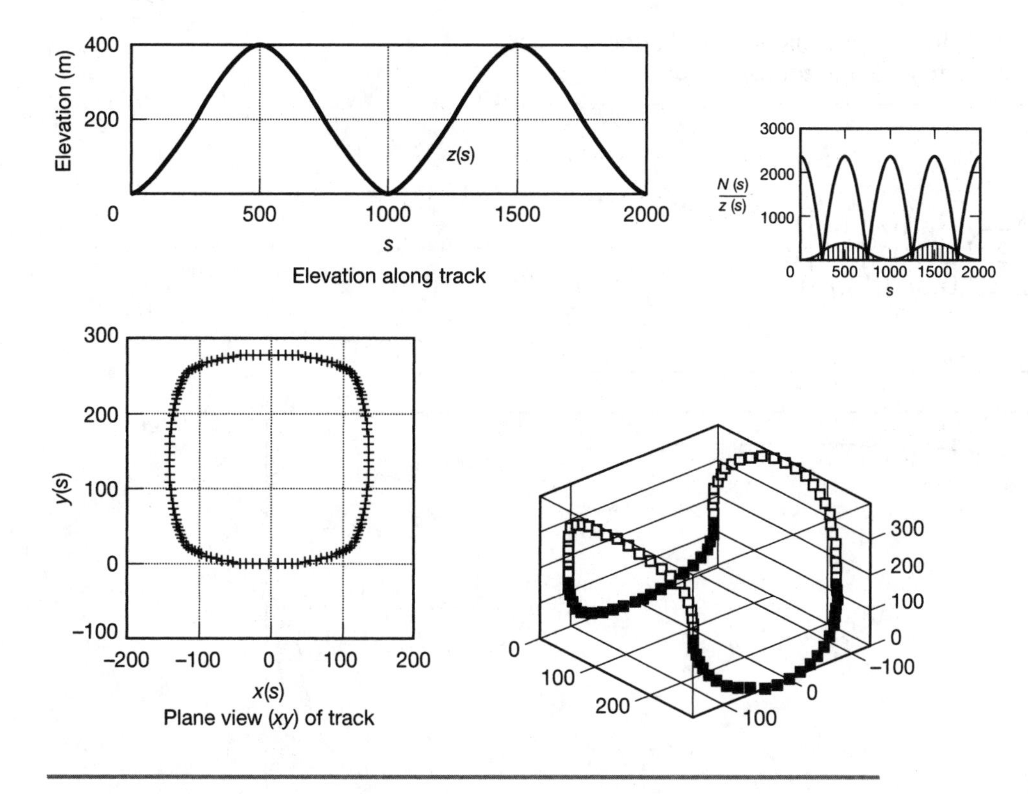

The solution to this sample problem requires the use of several functions and a `for` loop to compute successive values of a function defined by an integral, as was the case for the solution to Sample Problem 2.14. First, the three functions to be integrated are defined in m-files to be called later. These m-files are as follows:

```
function x=fx22(s);
theta=pi*s/1000;beta=(pi/2)*sin(pi*s/500);
x=cos(theta).*cos(beta);
```

```
function y=fy22(s);
theta=pi*s/1000;beta=(pi/2)*sin(pi*s/500);
y=sin(theta).*cos(beta);
```

```
function z=fz22(s);
beta=(pi/2)*sin(pi*s/500);
z=sin(beta);
```

Now, an m-file is created to call these three functions and integrate them over succesively larger values of s to form the coordinates:

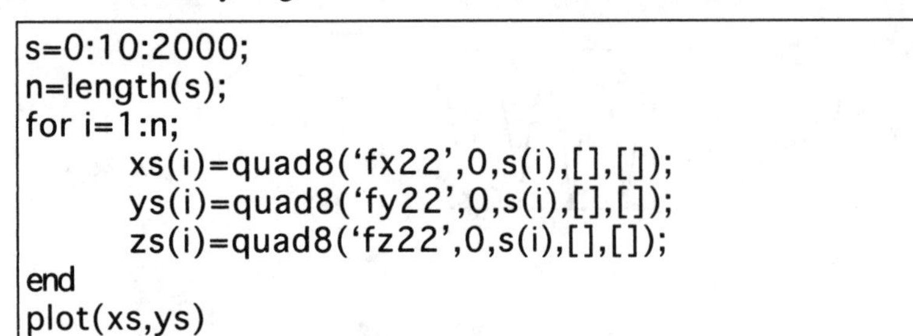

```
s=0:10:2000;
n=length(s);
for i=1:n;
        xs(i)=quad8('fx22',0,s(i),[],[]);
        ys(i)=quad8('fy22',0,s(i),[],[]);
        zs(i)=quad8('fz22',0,s(i),[],[]);
end
plot(xs,ys)
```

Next, the following command is typed in the command window to plot the z coordinate:

```
EDU>plot(s,zs),title('z(s) versus s')
```

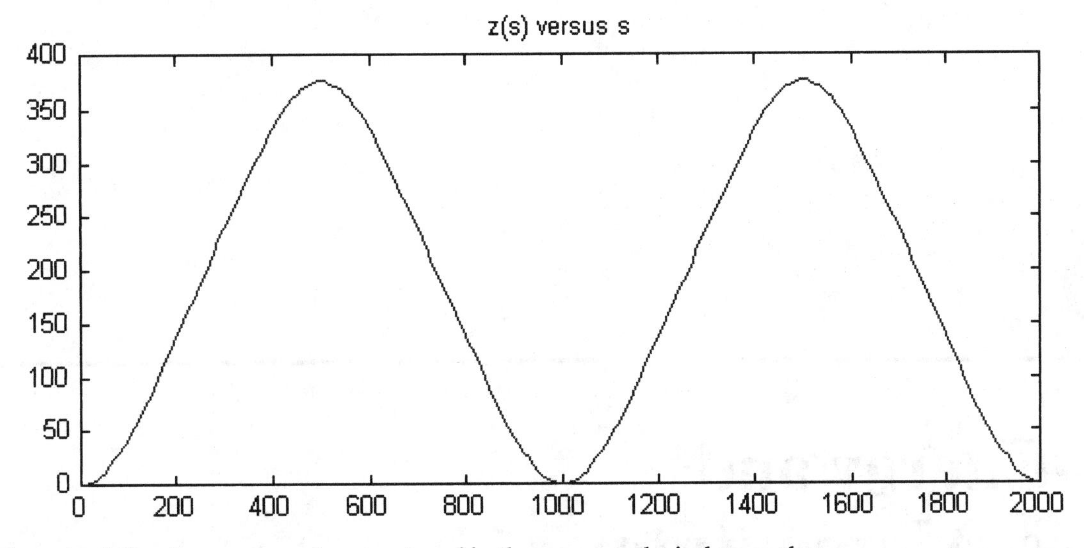

Next, the following commands are entered in the command window so that the normal force in Newtons is computed and plotted along with the z coordinate:

```
EDU>dt=pi/1000;db=(pi^2/1000)*cos(pi*s/500);
EDU>Ns=sqrt(dt^2*cos((pi/2)*sin(pi*s/500)).^2+db.^2)*15^2*1000;
EDU>plot(s,Ns,'+',s,zs,'*'),title('N(s), denoted *, and z(s),+, versus s')
```

These commands result in the following plot:

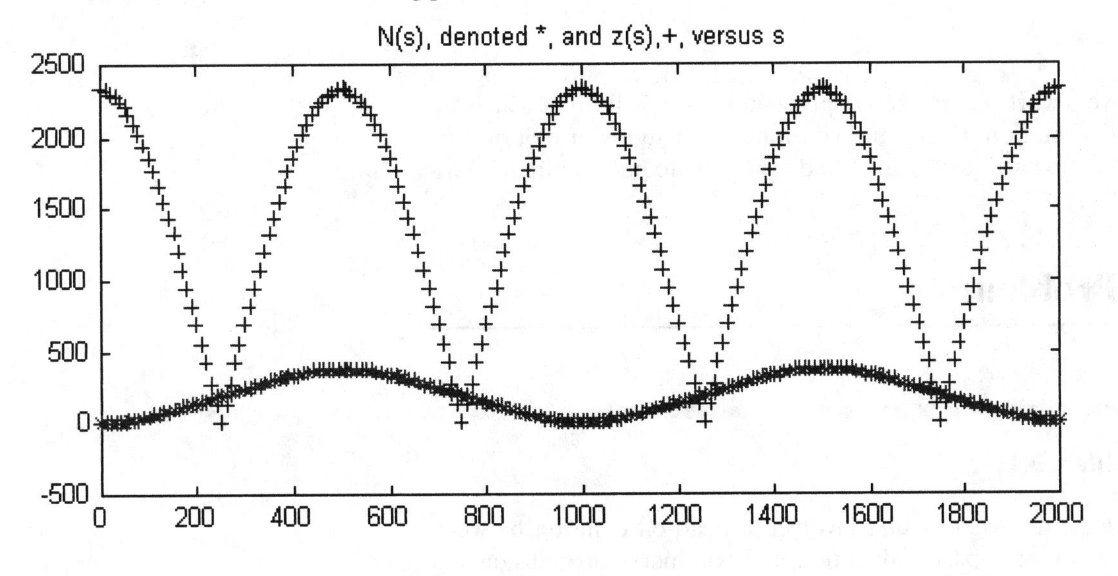

3

Work–Energy and Impulse–Momentum: First Integrals of Motion

Chapter 3 covers work–energy and impulse–momentum, both of which are first integrals of motion. Hence, the differential equations of motion that result in this section are easily solved and generally do not require numerical solutions.

Impact Problem

Sample Problem 3.11

Police investigate an accident that involved a head-on collision between two cars. The point of impact is identified, and skid marks precollision and postcollision are measured for each vehicle. The police diagram of the scene is shown in the accompanying diagram. The weight of vehicle A is 4000 lb, and the weight of vehicle B is 3200 lb. Tests on the pavement show that the coefficient of kinetic friction is 0.7 for both vehicles, and examination of the damage leads to an estimate of a coefficient of restitution as 0.5. As an accident reconstructionist, determine the initial velocities of the two vehicles.

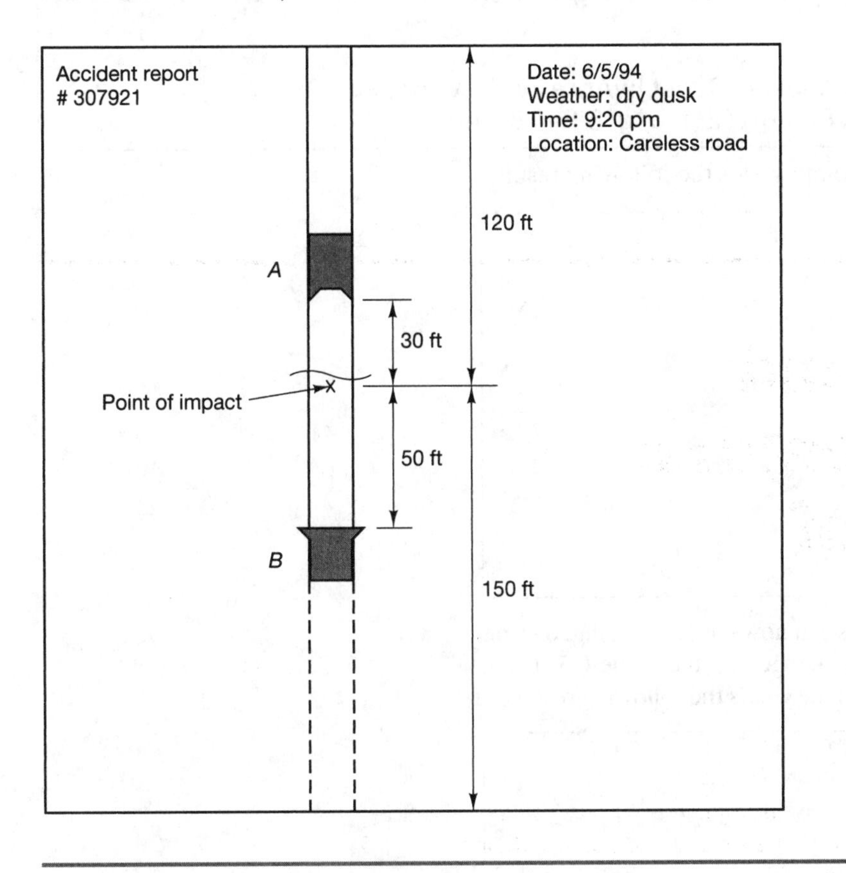

The accident-reconstruction problem shown in Sample Problem 3.7 can be developed into a general solution by using MATLAB. Let "Si" be the matrix for initial skid distance and "Sf" denote the final slide from the point of impact. The masses of the vehicles are designated by "mA" and "mB," and "mu" denotes the coefficient of kinetic friction between the tires and the pavement. In addition, "e" denotes the coefficient of restitution. The remaining symbols used in this solution should be obvious from the *Dynamics* text. The next set of commands can be used to compute "vAi" and "vBi" from the given information. These commands are entered into an m-file so that various parameters can be changed and the calculations remade. For instance, it is instructive to change the coefficient of friction for various weather conditions.

```
SiA=120;SiB=150;SfA=30;SfB=50;
e=0.5;mu=0.7;g=32.2;WA=4000;WB=3600;
vApost=sqrt(2*mu*g)*sqrt(SfA),vBpost=sqrt(2*mu*g)*sqrt(SfB),
syms vApre vBpre
V=[WA -WB;e e]^-1
      *[-WA*vApost+WB*vBpost;vApost+vBpost];vApre=V(1),vBpre=V(2)
vAi=sqrt(2*mu*g*SiA+vApre^2),vBi=sqrt(2*mu*g*SiB+vBpre^2)
```

Typing the name of this m-file after the prompt yields the following results:

```
vApost =
   36.7750
vBpost =
   47.4763
vApre =
   82.9505
vBpre =
   85.5521
vAi =
   110.8620
vBi =
   118.6640
```

As previously mentioned, the conditions can now easily be changed using the m-file. In particular, suppose that μ is changed to the value 0.5. Then, changing mu in the m-file and executing the file yields the following results:

```
EDU>impact
vApost =
   31.0805
vBpost =
   40.1248
vApre =
   70.1060
vBpre =
   72.3047
vAi =
   93.6955
vBi =
  100.2894
```

Next, consider the case for which $\mu = 0.7$ and the coefficient of restitution $e = 0.8$. As the coefficient of restitution is increased, less energy is lost during the impact, and since the energy after impact is known, a higher coefficient of restitution results in lower initial kinetic energy and, therefore, lower velocities. This set of conditions can be seen by changing mu and e in the previous m-file and then typing impact after the prompt in the command window:

```
EDU>impact
vApost =
   36.7750
vBpost =
   47.4763
vApre =
   53.0192
vBpre =
   52.2950
vAi =
   90.6677
vBi =
   97.4513
```

The solution to this sample problem provides an excellent example of the performance of a "what if" study. That is, by placing all of the calculations in an m-file, you can easily run the file again to see what happens if you change one of the coefficients. This method can be very useful for attempting to discover the cause of an event (e.g., changing one car's initial velocity in an accident reconstruction to see if the speed limit was exceeded) or for performing design studies.

4

Systems of Particles

Chapter 4 of *Dynamics* introduces the dynamics of systems of particles and lays the foundation for all of the rigid-body dynamics in subsequent chapters. MATLAB should be used in conjunction with this chapter only when it reduces the computational burden of long numerical calculations, as illustrated in the solution for Sample Problem 4.2. You may also use MATLAB to perform "what if" studies and to reduce calculational errors.

Sample Problem 4.2

A 2-kg particle explodes and breaks into three equal fragments when it has a velocity of $\mathbf{V} = 10\hat{\mathbf{i}}$ m/s. If the fragments are seen to travel in the following directions, relative to the particle after the explosion, determine the velocity of each fragment and the energy lost in the explosion:

$$\mathbf{V_a} = V_a\,(0.577\,\hat{\mathbf{i}} + 0.577\,\hat{\mathbf{j}} + 0.577\,\hat{\mathbf{k}})$$
$$\mathbf{V_b} = V_b\,(0.333\,\hat{\mathbf{i}} + 0.667\,\hat{\mathbf{j}} - 0.667\,\hat{\mathbf{k}})$$
$$\mathbf{V_c} = V_c\,(-0.667\,\hat{\mathbf{i}} + 0.667\,\hat{\mathbf{j}} + 0.333\,\hat{\mathbf{k}}).$$

To begin with, the data and calculations are entered into an m-file named fourpt2.m:

```
M=2;m=M/3;
ea=[0.577;0.577;0.577];eb=[0.333;0.667;-0.667];ec=[-0.667;0.667;0.333];
C=[m*ea m*eb m*ec];Li=[10*M;0;0];
V=inv(C)*Li; Va=V(1),Vb=V(2),Vc=V(3)
```

Then, fourpt2 is typed after the prompt to produce the following answers:

```
Va =
   20.7910
Vb =
    6.0072
Vc =
  -23.9928
```

This file can be easily altered to change M or any of the other parameters.

5

Kinematics of Rigid Bodies

MATLAB can be used in conjunction with numerous problems in Chapter 5 of *Dynamics* either to reduce the time it takes to work through the problems and to increase the accuracy of vector calculations or to help visualize the motion by graphing the displacement, velocity, or acceleration. The next sample problem illustrates some of these uses.

Sample Problem 5.8

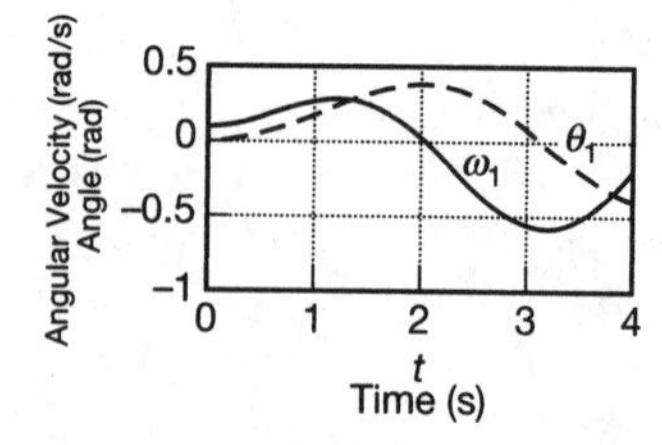

The differential equation of motion for a disk rotating about its axis is given by

$$\alpha + 0.2\,\omega^2 + 4\theta = \sin(t).$$

Determine the angular velocity and angular position as functions of time if the initial conditions are $\theta_0 = 0$ and $\omega_0 = 0.1$.

The solution to this problem uses an m-file to solve the differential equation of motion, as was done in the solution to Sample Problem 1.4. In this m-file, the respective time histories of angular position and velocity are stored in the vectors $x(1) = \omega$ and $x(2) = \theta$ so that $x(1)$ is the derivative of $x(2)$.

```
function xdot=fivept8(t,x);
xdot=[sin(t)-0.2*x(1)^2-4*x(2);x(1)];
```

Entering the following commands in the command window solves and plots the solution for both the angular position and velocity:

```
EDU>tspan=[0 4];
EDU>x0=[0.1;0];
EDU>ode45('fivept8',tspan,x0)
```

In the following plot, the curve that starts at zero is the angular displacement in radians, the other is the angular velocity radians per second, and both are plotted versus time in seconds:

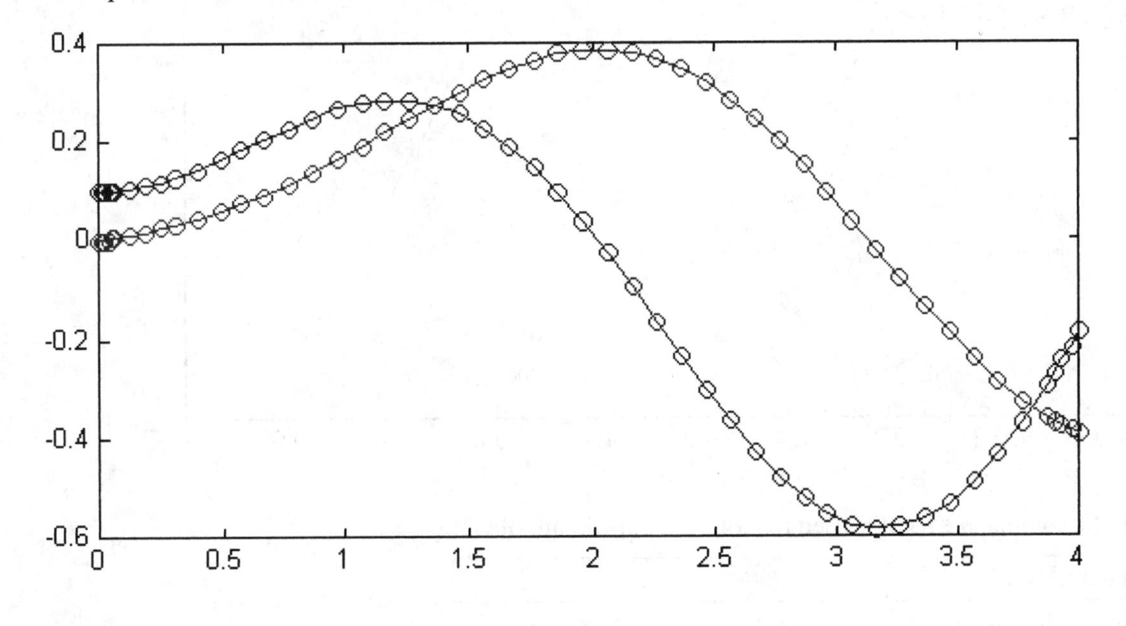

Sample Problem 5.14

Consider the automobile wheel assembly illustrated in the accompanying figure. If the wheel rolls without slipping and the velocity of the car is v to the right at the instant shown in the figure, determine the velocity of point B on the wheel. Leave the solution in terms of the radius r, the angle θ, and the velocity v.

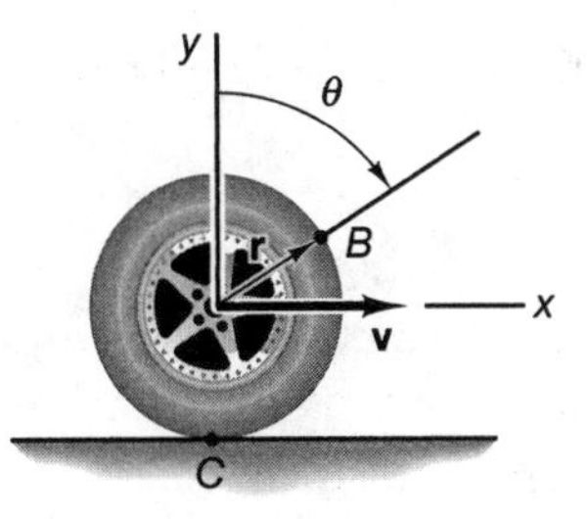

This problem is solved by using the symbolic code to form the vector calculations, and then the ezplot command is used to display the results:

```
EDU>syms t                          % define t as a symbolic quantity
EDU>v0=[100;0;0];w0=10;w=[0;0;-10];      % define the various constants
EDU>r=[10*sin(w0*t);10*(1+cos(w0*t));0];  % form r(t) symboliclly
EDU>vD=cross(w,r)+v0;              % form vD symbolically
EDU>vDx=vD(1);vDy=vD(2);NvD=sqrt(dot(vD,vD));   % define the coordinates and norm to be
                                                 % plotted
2EDU>ezplot(vDx,[0 1])             % plot the x coordinate of vD vs time
```

The previous list of commands produces the following plot of the x coordinate of vD versus time:

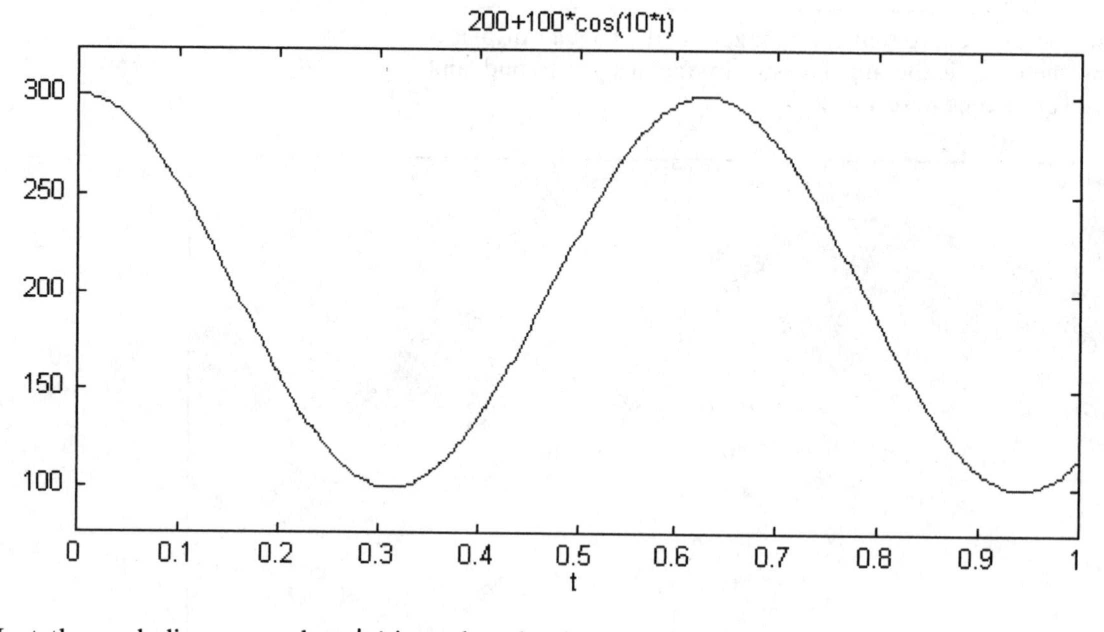

Next, the symbolic command ezplot is used to plot the y coordinate of vD:

```
EDU>ezplot(vDy,[0 1])
```

This command produces the following plot:

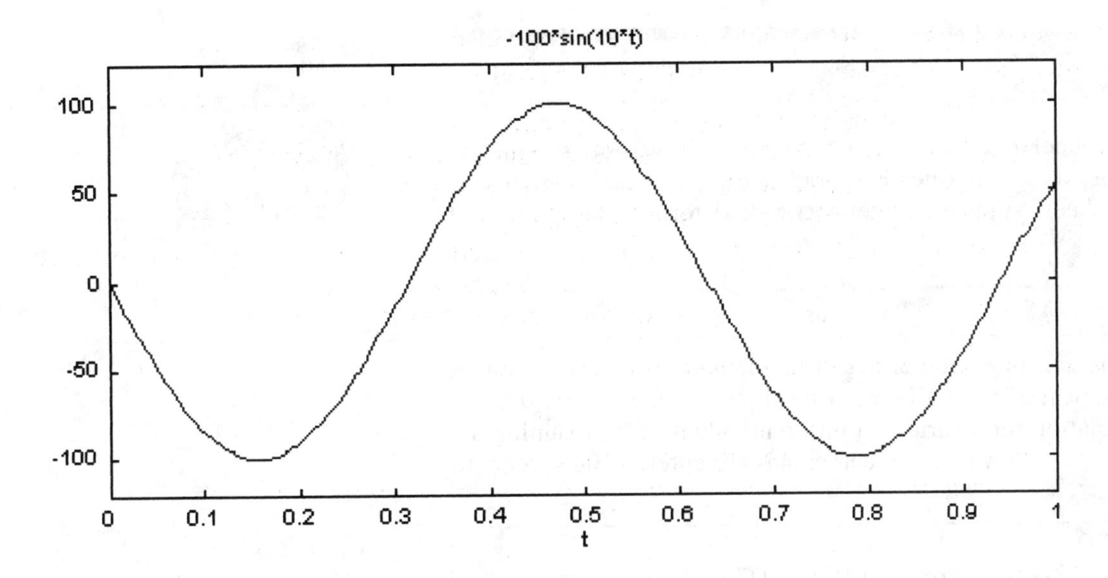

Next, the symbolic command **ezplot** is used to plot the norm of the vector vD versus time:

```
EDU>ezplot(NvD,[0 1])
```

This command produces the following plot:

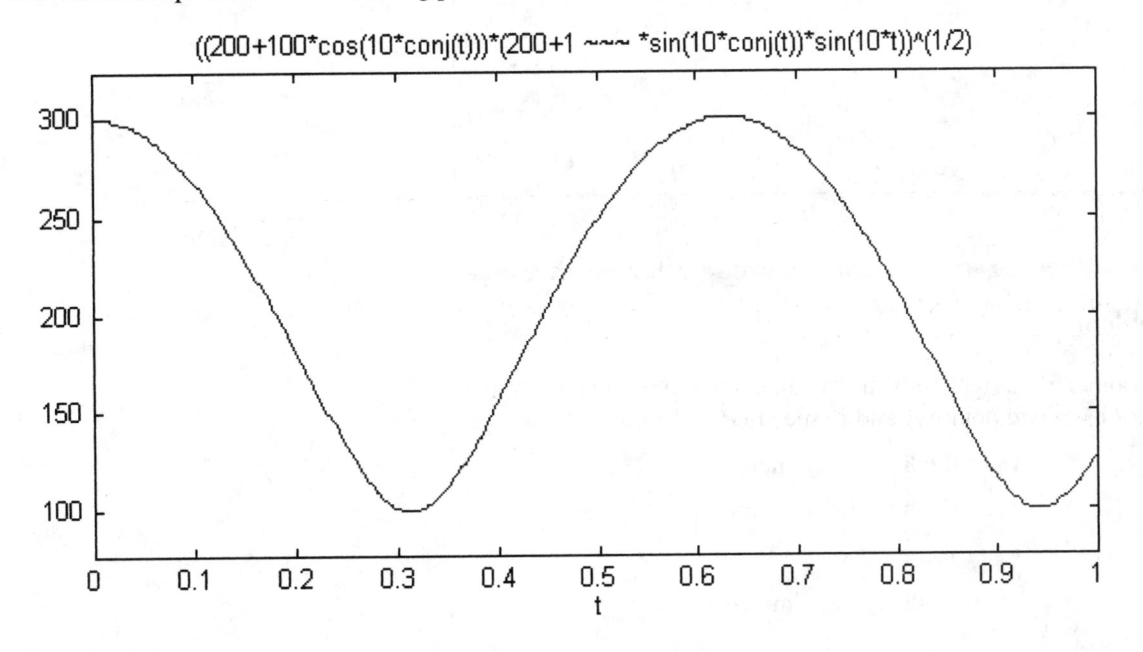

Sample Problem 5.15

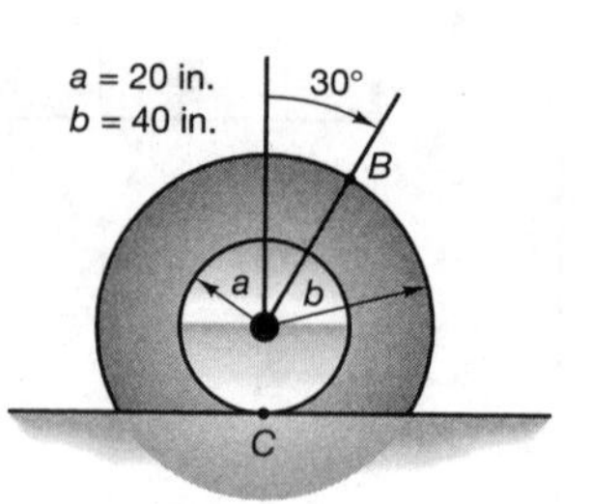

The compound wheel in the accompanying figure rolls without slipping on the inner hub of the wheel, which has a radius of 20 inches. If the angular velocity of the wheel is 4 rad/s counterclockwise, determine the velocity of point *B*.

This problem can be solved either in the command window or in an m-file. However, if it is written in an m-file, then the file can be used to repeat the calculation for a variety of different values of θ. Assuming the use of an m-file, the following set of commands are entered and saved into the m-file:

```
th=30*pi/180;
r=[40*sin(th);20+40*cos(th);0];w=[0;0;4];
vb=cross(w,r)
normvb=sqrt(dot(vb,vb))
```

This m-file produces the following values:

```
EDU>fivept15
vb =
 -218.5641
   80.0000
         0
normvb =
  232.7450
```

Sample Problem 5.16

In Sample Problem 5.9, a rigid body moving in plane motion had the velocities and positions of two points, *A* and *B*, specified as follows:

$$\mathbf{r}_A = 1.60\hat{\mathbf{i}} + 1.50\hat{\mathbf{j}} \text{ meter}$$
$$\mathbf{r}_B = 2.00\hat{\mathbf{i}} + 1.80\hat{\mathbf{j}} \text{ meter}$$
$$\mathbf{v}_A = 3.00\hat{\mathbf{i}} \text{ m/sec}$$
$$\mathbf{v}_B = 2.40\hat{\mathbf{i}} + 0.80\hat{\mathbf{j}} \text{ m/sec.}$$

We found that the angular velocity is equal to 2 rad/s at the instant of time considered in the sample problem. If the linear accelerations of points A and B are

$$\mathbf{a}_A = 3\hat{\mathbf{i}} + 2\hat{\mathbf{j}} \text{ m/s}^2$$

$$\mathbf{a}_B = 1.1\hat{\mathbf{i}} + 1.2\hat{\mathbf{j}} \text{ m/s}^2$$

determine the angular acceleration of the rigid body at this instant.

The solution to this problem involves the simple use of MATLAB as a vector calculator. The first two lines of code in the next box enter the relevant position, velocity, and acceleration vectors. The third line computes the relative position, velocity, and acceleration. The remaining lines compute and display ω, α, and $\mathbf{r}_{CA}$, respectively.

```
EDU>rB=[2;1.8;0];rA=[1.6;1.5;0];vA=[3;0;0];vB=[2.4;0.8;0];
EDU>aA=[3;2;0];aB=[1.1;1.2;0];
EDU>rBA=rB-rA;vBA=vB-vA;aBA=aB-aA;
EDU>w=cross(rBA,vBA/dot(rBA,rBA))
w =
     0
     0
 2.0000
EDU>alpha=cross(rBA,aBA/dot(rBA,rBA))
alpha =
     0
     0
 1.0000
EDU>rCA=cross(w,vA/dot(w,w))
rCA =
     0
 1.5000
     0
```

Section 5.10: Analysis of Plane Motion in Terms of a Parameter

For this section of the *Dynamics* text, the use of MATLAB as a graphical calculator allows the visualization of the angular position and velocity of the sliding rod, as illustrated in the next list of commands. The symbolic code is used so that the derivative can be taken analytically.

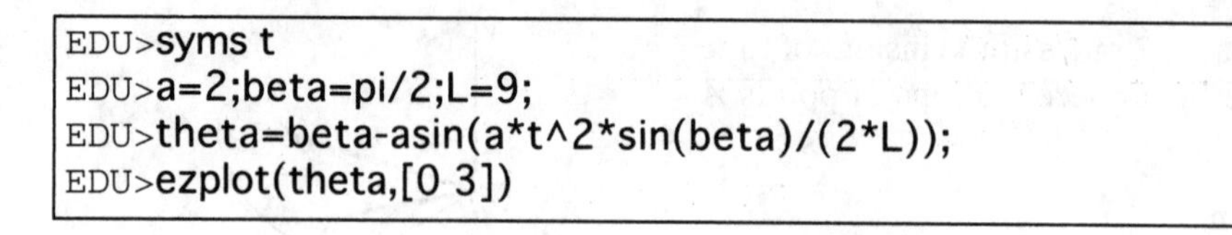

```
EDU>syms t
EDU>a=2;beta=pi/2;L=9;
EDU>theta=beta-asin(a*t^2*sin(beta)/(2*L));
EDU>ezplot(theta,[0 3])
```

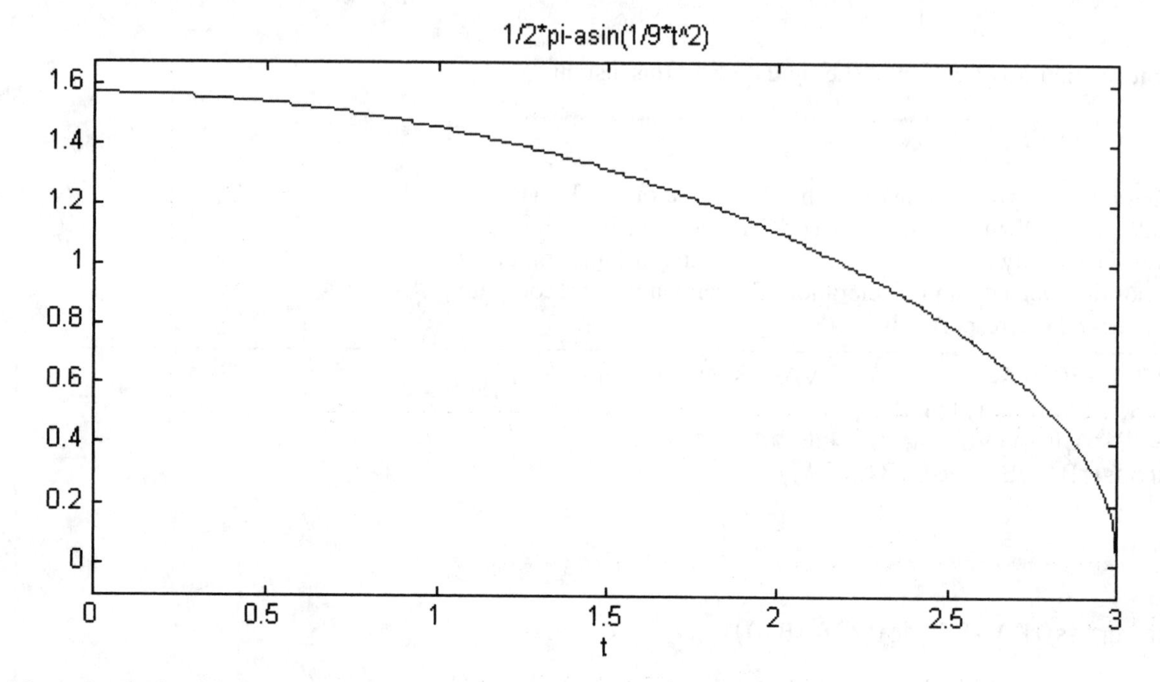

Next, the derivative is computed symbolically and plotted by the following commands:

```
EDU>omega=diff(theta,t);
EDU>ezplot(omega,[0 3])
```

These commands produce the following plot of the angular velocity versus time:

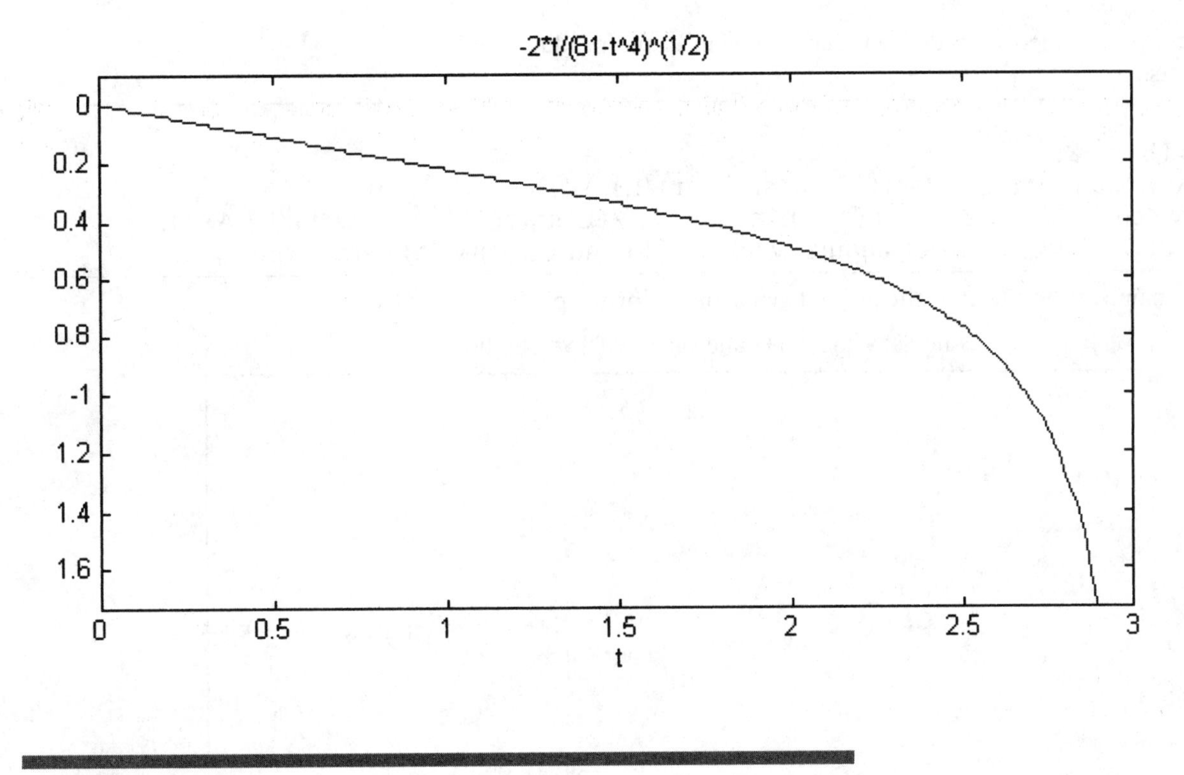

Sample Problem 5.21

Develop a general solution for the motion of the piston shown in the accompanying diagram if the disk is rotating counterclockwise at a constant rate ω.

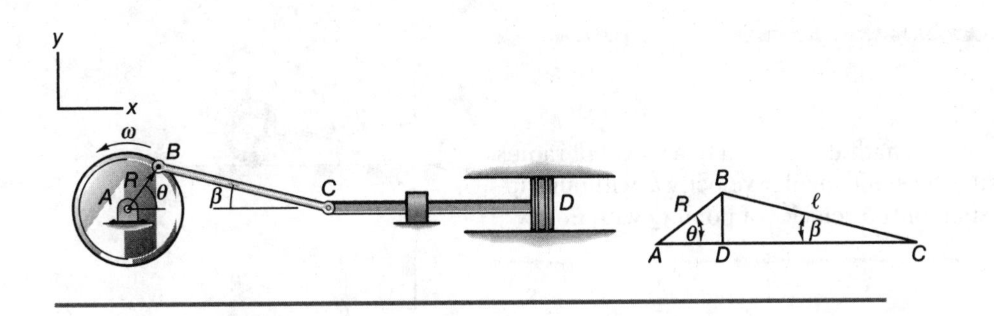

The MATLAB solution to this sample problem simply plots the solution of the piston problem numerically. An m-file is used because it allows the parameters to be varied if so desired. The following m-file produces the desired plots:

```
t=0:0.1:5;
R=6; L=18; w=2;
wBC=-R*w*cos(w*t)./(L*sqrt(1-(R*sin(w*t)/L).^2));
vC=-R*w*sin(w*t)-R^2*w*sin(w*t).*cos(w*t)./(L*sqrt(1-(R*sin(w*t)/L).^2));
plot(t,wBC,'+',t,vC,'*'),title('angular velocity (+) and velocity (*) versus time')
```

Typing the name of this file after the prompt yields the following plot:

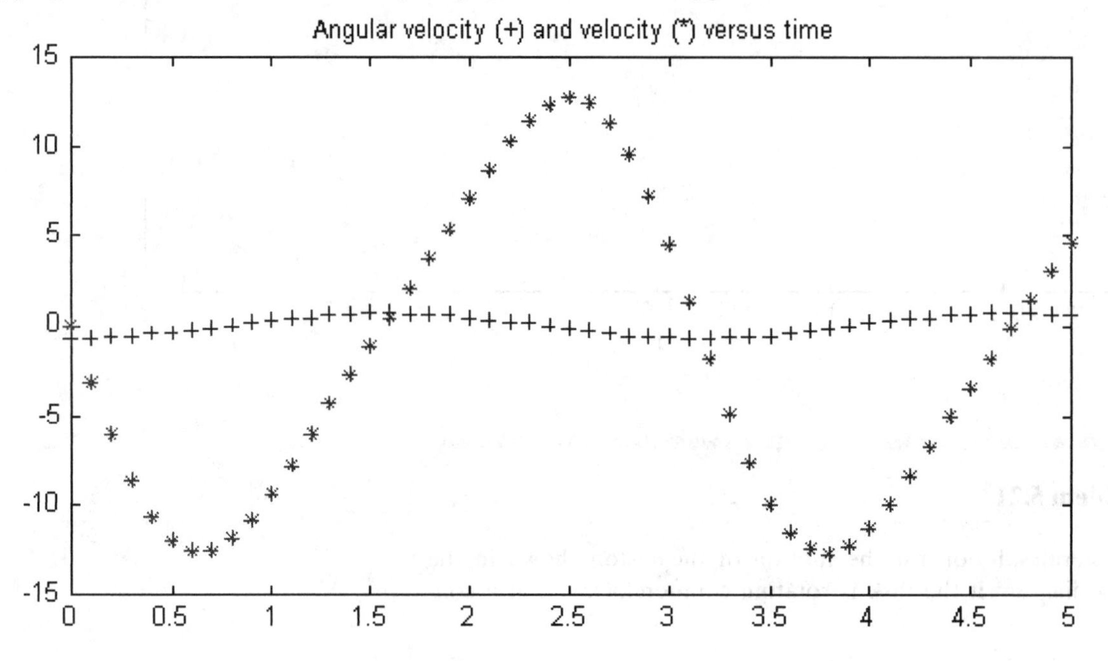

Sample Problem 5.23

In the accompanying figure, a rod of length L is pinned to a wheel of radius r at point P. If the wheel rolls at a constant angular velocity ω without slipping, develop a general expression for the velocity of point Q with time.

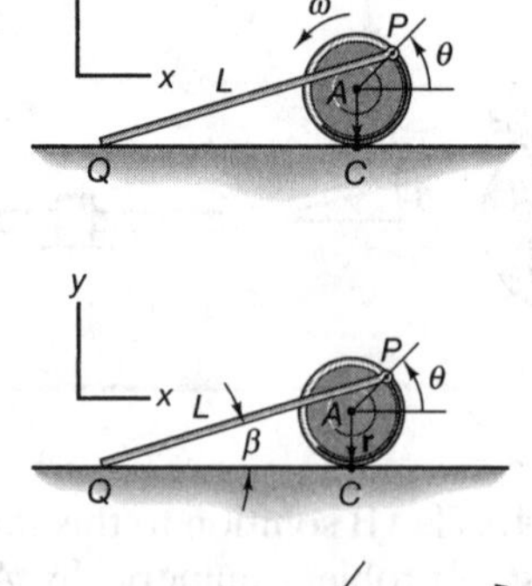

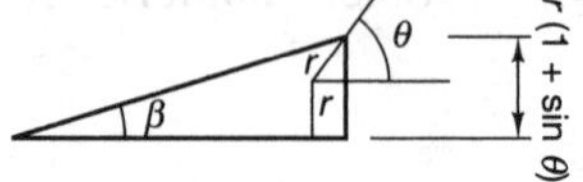

This example of a rolling wheel with an attached rod is easily computed in MATLAB. If the commands are put in an m-file, the parameters can easily be changed. Note, however, that the length of the rod must always be chosen to be greater than the diameter of the wheel in order to keep the end of the rod in contact with the ground. The following m-file computes and plots the angle β and the velocity of the point Q:

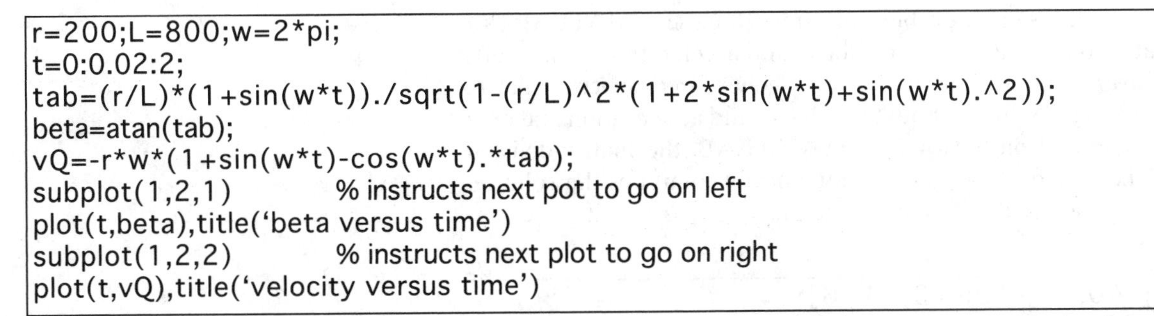

```
r=200;L=800;w=2*pi;
t=0:0.02:2;
tab=(r/L)*(1+sin(w*t))./sqrt(1-(r/L)^2*(1+2*sin(w*t)+sin(w*t).^2));
beta=atan(tab);
vQ=-r*w*(1+sin(w*t)-cos(w*t).*tab);
subplot(1,2,1)          % instructs next pot to go on left
plot(t,beta),title('beta versus time')
subplot(1,2,2)          % instructs next plot to go on right
plot(t,vQ),title('velocity versus time')
```

Typing the name of this file after the prompt produces the following plots:

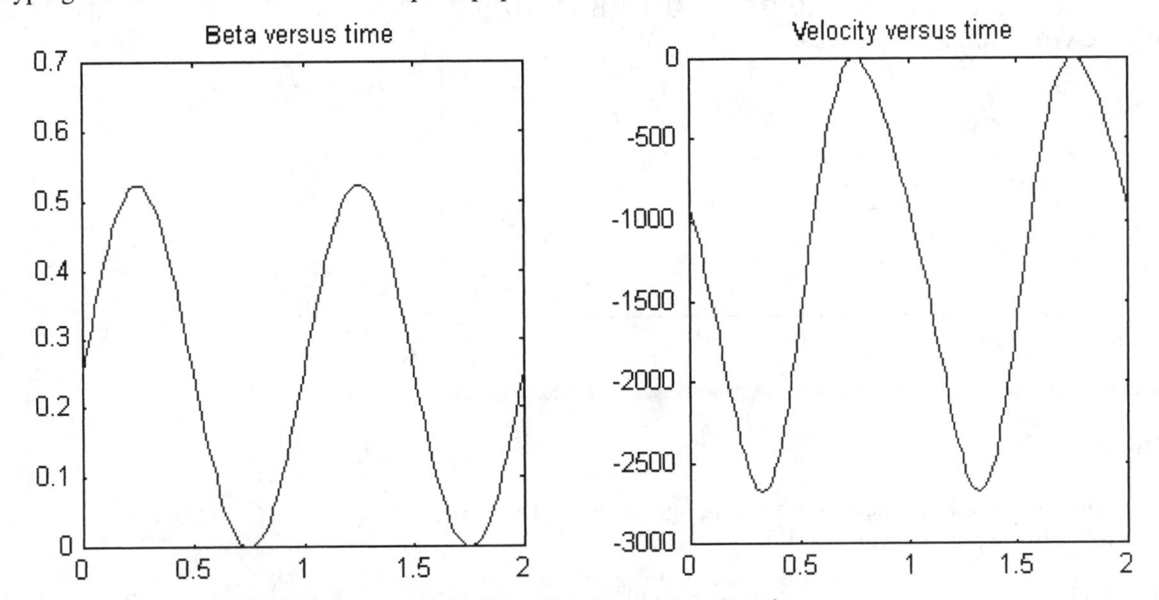

Note the use of the subplot command to allow the plots to be placed next to each other. Subplot(n,m,p) splits the page into $n \times m$ parts and makes the line following the "subplot" command print in part p of the screen, where p is quadrant 1, 2, 3, or 4 of the screen, numbered clockwise starting in the upper left corner of the screen.

Sample Problem 5.24

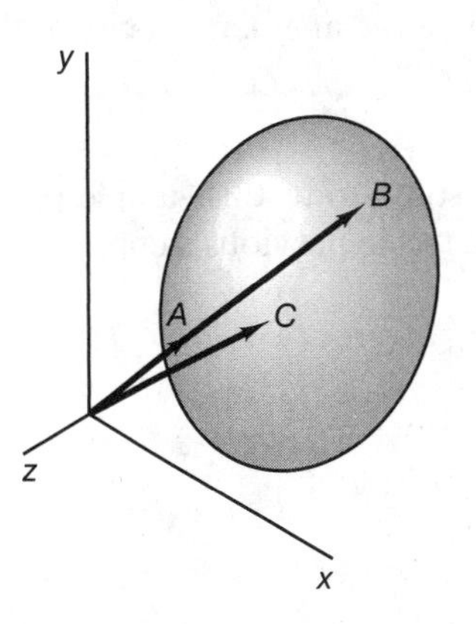

The position, linear velocities and accelerations of three noncollinear points on a rigid body are given in the table below:

	***r** mm*			***v** mm/s*			***a** mm/s²*		
	y	*z*	*x*	*y*	*z*	*x*	*y*	*z*	
A	100	100	0	600	−400	100	850	1200	−240
B	300	300	0	200	0	0	200	200	0
C	220	180	0	440	−160	40	420	760	−140

Determine the angular velocity and the angular acceleration of the body.

The solution to this sample problem illustrates the use of MATLAB as a vector calculator to solve for angular velocity and acceleration using a left-generalized inverse matrix. Note that the generalized inverse matrix $Cd = (C^TC)^{-1}C^T$ is the same for both the velocity and acceleration and only needs to be computed once. However, in MATLAB, the matrix-division operator ("\"; i.e., the backslash key) automatically computes the solution that minimizes the squared error in the difference $A\mathbf{x} - \mathbf{b}$, which is the generalized inverse.

```
EDU>vrel=[-40;40;-10;-16;24;-6];
EDU>qrel=[20;-20;4;8;-12;2];
EDU>C=[0 0 -20;0 0 20;20 -20 0;0 0 -8;0 0 12;8 -12 0];
EDU>omega=C\vrel,alpha=C\qrel
omega =
  -0.0000
   0.5000
   2.0000
alpha =
   0.1000
  -0.1000
  -1.0000
```

Sample Problem 5.25

Consider the rigid body, shown in the accompanying figure, having the position and velocity of three points given as follows:

$$\mathbf{r}_A = 2\hat{\mathbf{i}} - \hat{\mathbf{j}} + 3\hat{\mathbf{k}}\ \text{m} \qquad \mathbf{v}_A = 3\hat{\mathbf{i}} - 2\hat{\mathbf{j}} + \hat{\mathbf{k}}\ \text{m/s}$$

$$\mathbf{r}_B = 3\hat{\mathbf{j}} - \hat{\mathbf{k}}\ \text{m} \qquad \mathbf{v}_B = 19\hat{\mathbf{i}} + 10\hat{\mathbf{j}} + 5\hat{\mathbf{k}}\ \text{m/s}$$

$$\mathbf{r}_C = \hat{\mathbf{i}} + 2\hat{\mathbf{j}} - 2\hat{\mathbf{k}}\ \text{m} \qquad \mathbf{v}_C = 23\hat{\mathbf{i}} + 15\hat{\mathbf{j}} + 5\hat{\mathbf{k}}\ \text{m/s}.$$

Determine the angular velocity of the body.

z
A
B
C
x
y

The solution to this sample problem is identical to the code from the solution to the previous sample problem.

```
EDU>vrel=[16;12;4;20;15;5];
EDU>C=[0 -4 -4;4 0 -2;4 2 0;0 -5 -3;5 0 -1;3 1 0];
EDU>omega=C\vrel
omega =
   3.0000
  -4.0000
   0.0000
```

Sample Problem 5.26

The mechanism shown in the acompanying figure converts rotational motion into translational motion. Develop a relationship between the angular velocity of the wheel and the linear velocity of the collar on the shaft. The connections on bar AB at points A and B are both ball-and-socket connections.

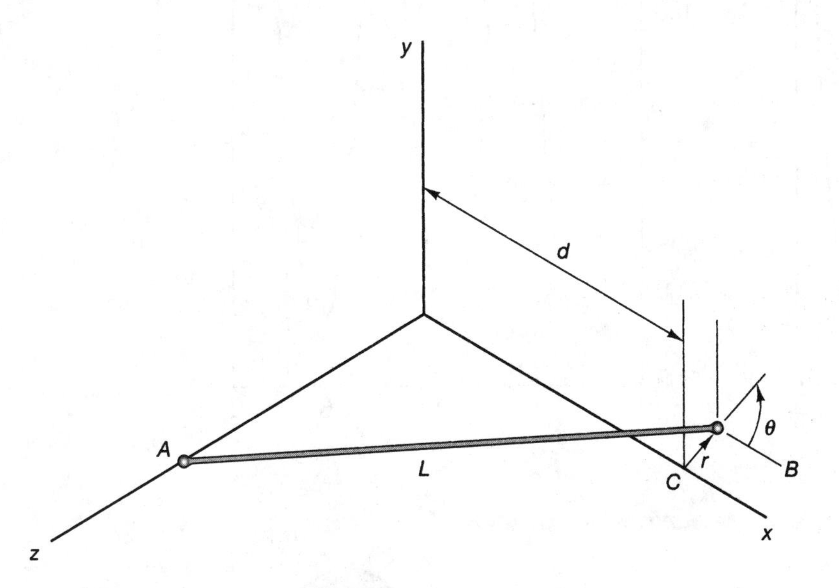

Let r be the distance from the disk center C to the socket B. Assume that the disk rotates with a constant angular velocity ω.

The velocity and acceleration can be plotted by entering the velocity symbolically and using the symbolic differentiation to compute the acceleration and the ezplot command to plot each versus time.

```
EDU>syms t
EDU>w=10;L=21;d=9;r=6;
EDU>v=r*d*sin(w*t)*w/sqrt(L^2-r^2-2*r*d*cos(w*t));
EDU>a=diff(v,t);
EDU>subplot(1,2,1),ezplot(v,[0 2])
EDU>subplot(1,2,2),ezplot(a,[0 2])
```

The previous set of commands produces the following two plots:

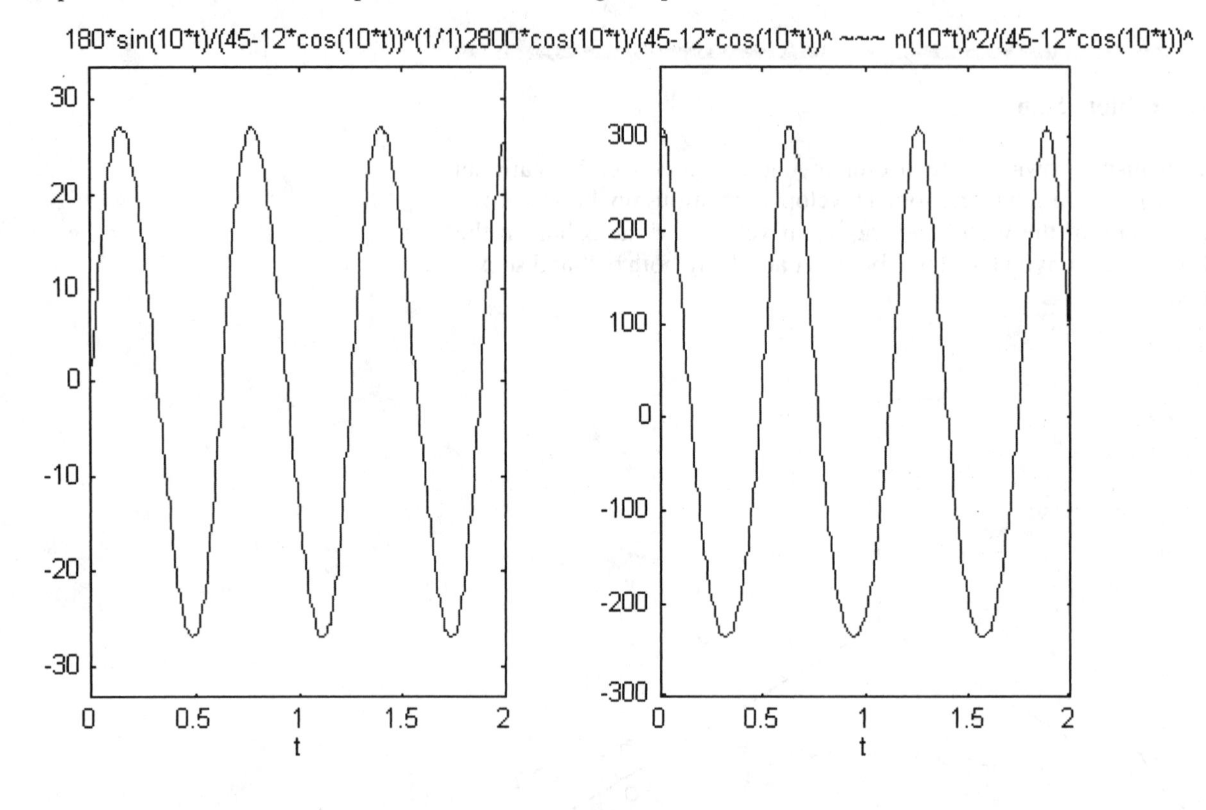

Sample Problem 5.31

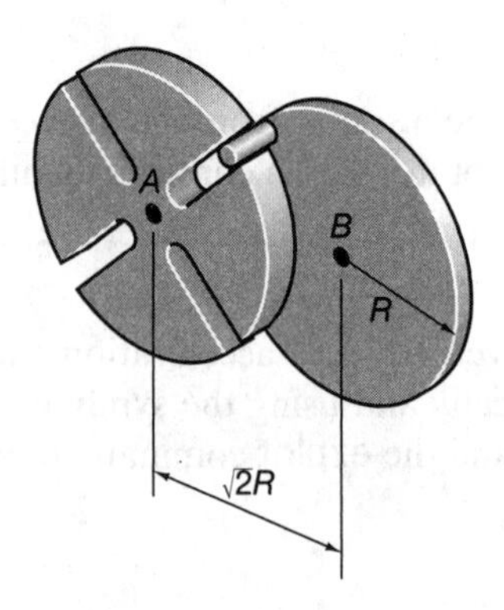

An intermittent motion device called the Geneva, or Malta, mechanism is shown in the accompanying figure. This mechanism allows intermittent rotation of disk A as disk B rotates at a constant rate. The Geneva wheel, disk A, is fitted with at least three equispaced, radial slots. Disk B has a pin that enters a radial slot and causes the Geneva wheel to turn through a portion of a revolution. When the pin leaves a slot, the Geneva wheel will remain stationary until the pin enters the next slot. A Geneva wheel needs a minimum of three slots to work, but the maximum number of slots is limited only by the size of the wheel. In this case, disk A will rotate $1/4$ turn for each full rotation of disk B. This information allows for a method of counting rotations and is useful for many machines. If disk B is rotating counterclockwise at a constant rate $\boldsymbol{\omega}$, determine the angular velocity and the angular acceleration of disk A when the pin P is engaged. Determine the velocity of the pin relative to the disk A during engagement.

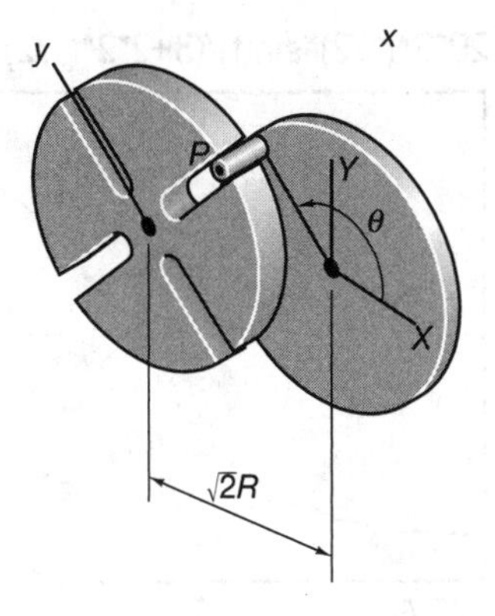

The code for the solution of this problem represents the solution of the Geneva Mechanism because the expressions for the velocity of the pin relative to the Geneva wheel and the angular velocity of the wheel are indicated. These quantities are then differentiated symbolically, and all four quantities are plotted. The following m-file is used to perform this task:

```
syms t
R=10;w=2;t1=0.75*pi;t2=1.25*pi;
vx=-R*sqrt(2)*sin(t)*w/sqrt(3+2*sqrt(2)*cos(t));
subplot(2,2,1),ezplot(vx,[t1 t2])
vb=w*(1+sqrt(2)*cos(t))/(3+2*sqrt(2)*cos(t));
subplot(2,2,2),ezplot(vb,[t1 t2])
ax=diff(vx,t)*w;
subplot(2,2,3),ezplot(ax,[t1 t2]),title('x accel vs theta')
ab=diff(vb,t)*w;
subplot(2,2,4),ezplot(ab,[t1 t2]),title('beta accel vs theta')
```

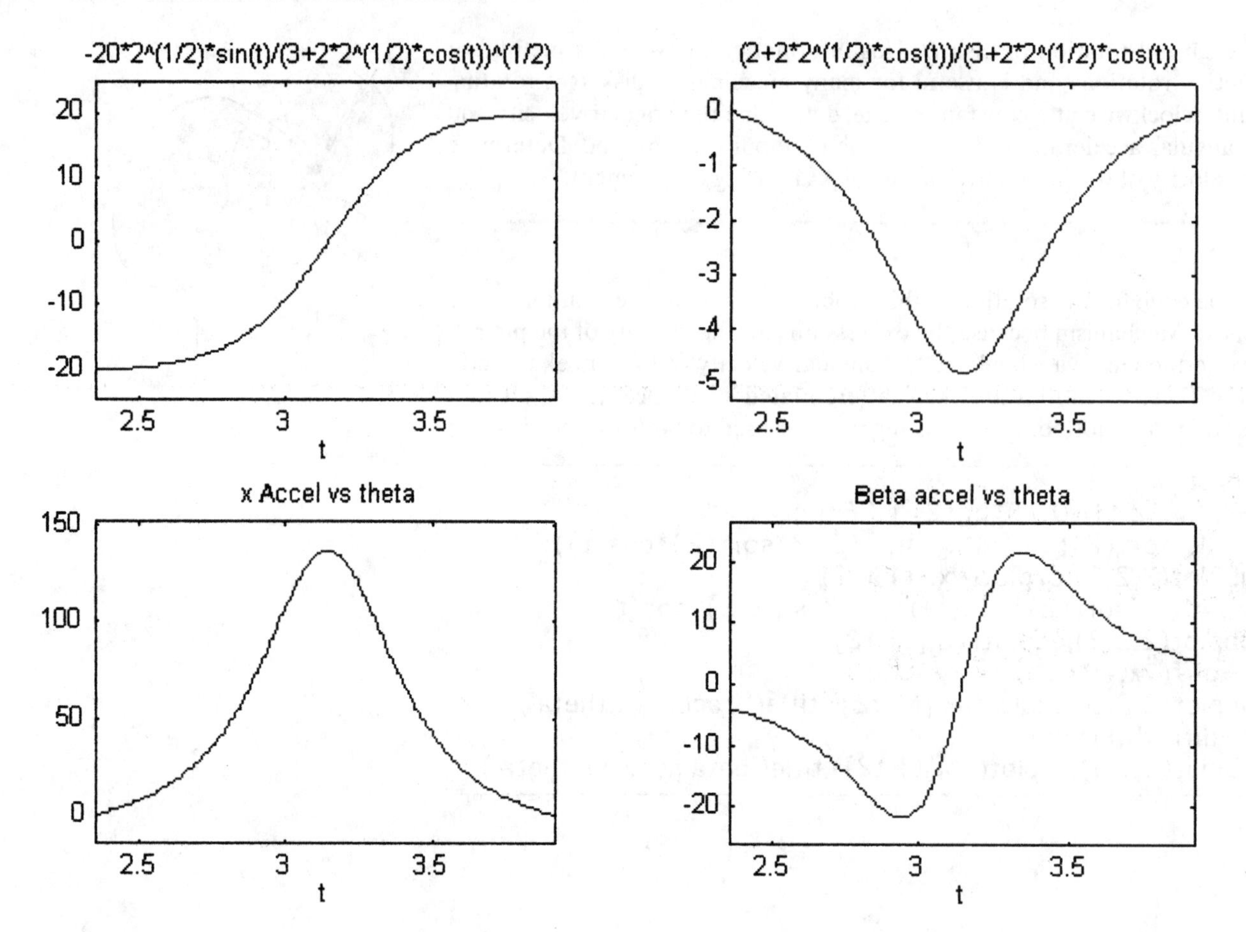
-20*2^(1/2)*sin(t)/(3+2*2^(1/2)*cos(t))^(1/2)
(2+2*2^(1/2)*cos(t))/(3+2*2^(1/2)*cos(t))
x Accel vs theta
Beta accel vs theta
t
t
t
t
20
10
0
-10
-20
0
-1
-2
-3
-4
-5
150
100
50
0
20
10
0
-10
-20
2.5
3
3.5

6

Dynamics of Rigid Bodies in Plane Motion

In this chapter of the supplement, MATLAB is used to assist in the solution of many problems. Initially, it is used to solve systems of linear equations, much as was done repeatedly in *Statics*. In these initial problems, the acceleration is a constant, and therefore the resulting differential equations can be solved analytically and numerical solutions of the equations of motion are not required. As we progress to more realistic and complex motions, many of the resulting differential equations will be nonlinear and will not be analytically solvable. Hence, the real significance of using MATLAB will come into play, as the computational code allows you to compute solutions to these nonlinear equations of motion. Before these codes were developed, these nonlinear problems were solved in a quasistatic manner—that is, solved only for the given position or instant of time.

You are encouraged to obtain complete solutions of the plane-dynamics problems and then to plot the results so that you develop a complete understanding of the motion and the effects of the various parameters on the solution of the equations. This method is essential for any practicing engineer involved in designing dynamics systems. Before the availability of PCs and programs such as MATLAB, many of these problems could not be solved or required the use of large codes and many hours of programming and computing time. As in all engineering fields, the computerization that has occurred over the last few years has radically changed the way in which engineers work. The skills you learn in dynamics and the integration of dynamics with MATLAB will help you a great deal in your profession following your engineering education.

The use of this code is intended not only to provide you with an ability to solve complex dynamics problems, but also to help you build intuition and design skills by "playing" with your solutions in terms of changing the initial conditions, varying parameters, and essentially asking "what if?" as you work through the problems. You are also encouraged to use the code to your best ability. The m-files and commands used here are on the spartan side, written with an attempt to be as brief and simplistic as possible. If you enjoy using the code, you will undoubtedly think of better ways to use it to understand dynamics than the way that we have presented in this supplement. You are also encouraged to seek more interesting uses of MATLAB.

Sample Problem 6.1: The Tipping Box

Consider an individual attempting to push a large box across the floor, as shown in the accompanying diagram. Determine the resulting movement of the box when its size and weight are known. Develop a general approach for any applied force, at any position, and for a particular coefficient of static and kinetic friction between the box and the floor.

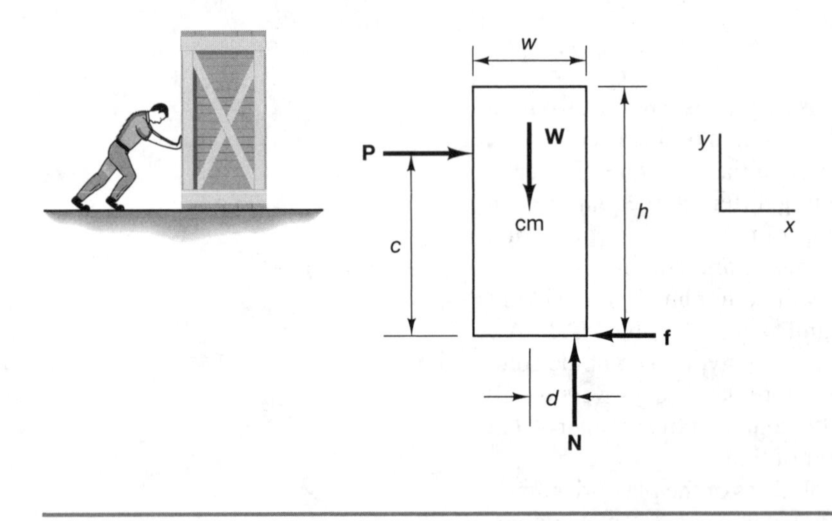

In the solution to this problem, a symbolic solution is used to avoid algebraic errors:

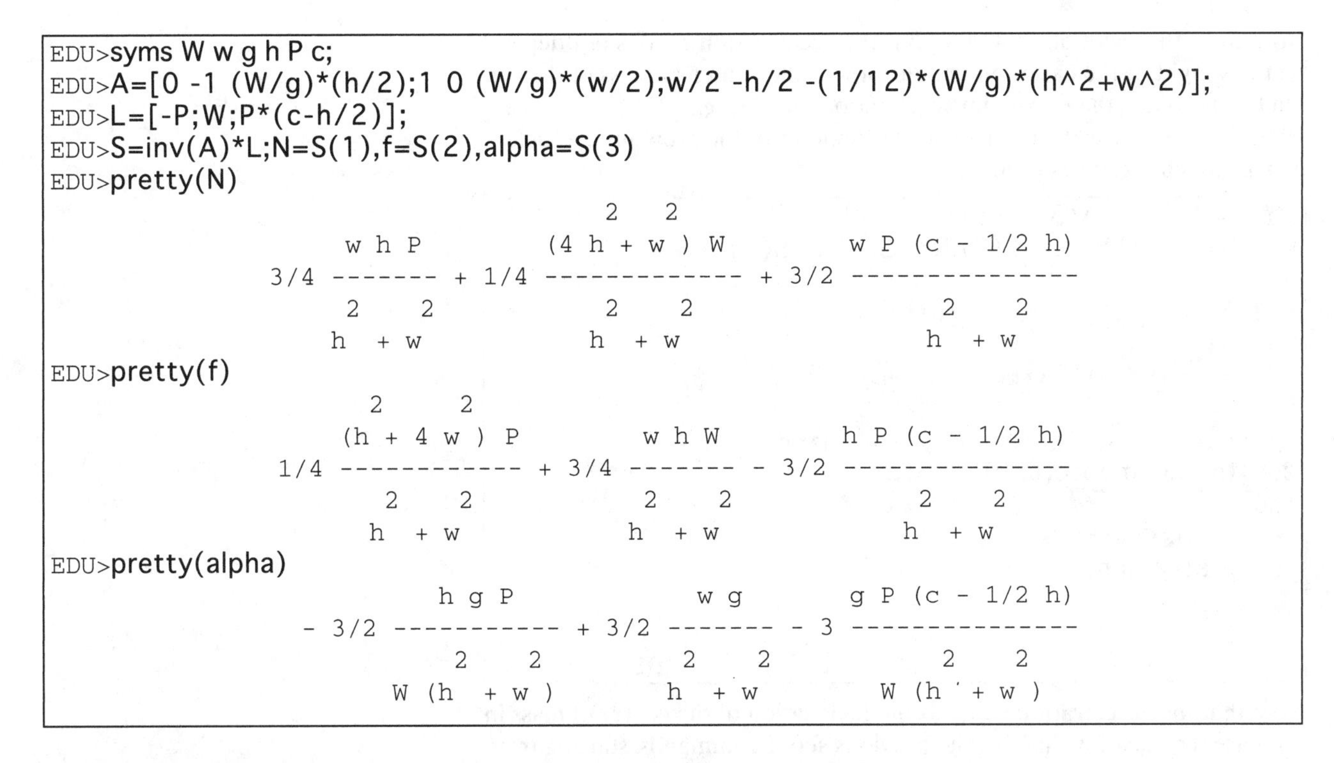

```
EDU>syms W w g h P c;
EDU>A=[0 -1 (W/g)*(h/2);1 0 (W/g)*(w/2);w/2 -h/2 -(1/12)*(W/g)*(h^2+w^2)];
EDU>L=[-P;W;P*(c-h/2)];
EDU>S=inv(A)*L;N=S(1),f=S(2),alpha=S(3)
EDU>pretty(N)
                                     2    2
              w h P            (4 h  + w ) W          w P (c - 1/2 h)
     3/4 ------- + 1/4 ------------- + 3/2 ---------------
              2    2               2    2                  2    2
             h  + w               h  + w                  h  + w
EDU>pretty(f)
                 2      2
             (h  + 4 w ) P           w h W          h P (c - 1/2 h)
     1/4 ------------ + 3/4 ------- - 3/2 ---------------
                 2    2              2    2               2    2
                h  + w              h  + w               h  + w
EDU>pretty(alpha)
                     h g P                  w g        g P (c - 1/2 h)
     - 3/2 ----------- + 3/2 ------- - 3 ---------------
                    2    2                  2    2                 2    2
               W (h  + w )              h  + w            W (h  + w )
```

Sample Problem 6.3

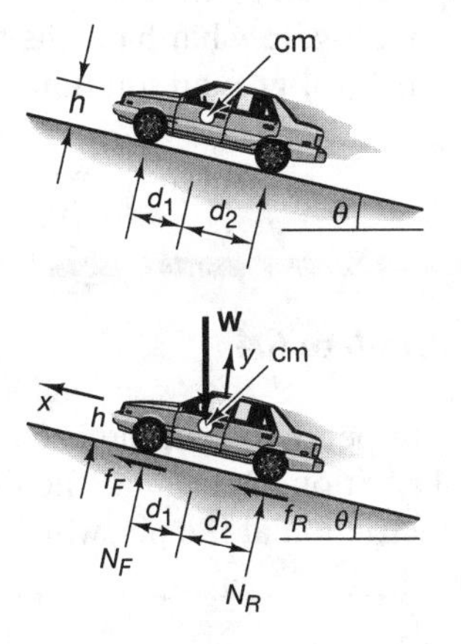

Determine the maximum acceleration that a car can obtain when climbing a hill of slope θ if the coefficient of static friction between the tires and the road is μ_S. The center of mass of the car is toward the front of the car, due to the weight of the engine. (See the accompanying diagram.) Determine the maximum acceleration if the car is a (a) four-wheel-drive, (b) rear-wheel-drive, and (c) front-wheel-drive vehicle.

Note that in the following MATLAB solution to this sample problem, "th" is used for θ and "mu" for μ_S.

```
EDU>syms mu d1 d2 h W g th
EDU>A=[mu mu -W/g;1 1 0;d1+mu*h -d2+mu*h 0];
EDU>L=W*[sin(th);cos(th);0];
EDU>S=inv(A)*L;Na=S(1),Nb=S(2),a=S(3)
Na =
(d2-mu*h)/(d2+d1)*W*cos(th)
Nb =
(d1+mu*h)/(d2+d1)*W*cos(th)
a =
-g*sin(th)+g*mu*cos(th)
```

Note from this symbolic solution that the acceleration a does not depend on the weight of the car or on the location of the center of mass of the car. On level ground, the maximum acceleration is $a = \mu_s g$.

In this next case, the (1,1) and (3,1) elements of the previous coefficient matrix are changed to obtain:

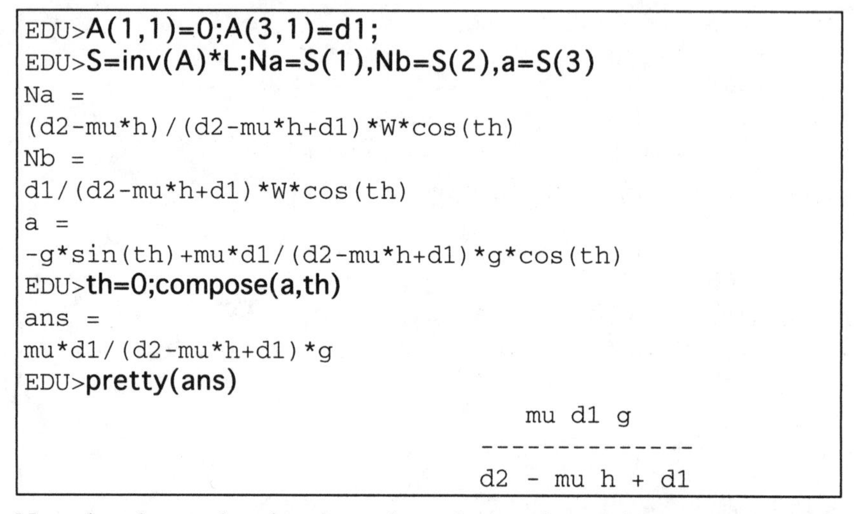

```
EDU>A(1,1)=0;A(3,1)=d1;
EDU>S=inv(A)*L;Na=S(1),Nb=S(2),a=S(3)
Na =
(d2-mu*h)/(d2-mu*h+d1)*W*cos(th)
Nb =
d1/(d2-mu*h+d1)*W*cos(th)
a =
-g*sin(th)+mu*d1/(d2-mu*h+d1)*g*cos(th)
EDU>th=0;compose(a,th)
ans =
mu*d1/(d2-mu*h+d1)*g
EDU>pretty(ans)
                                    mu d1 g
                                 --------------
                                 d2 - mu h + d1
```

Note that the acceleration depends on the location of the center of mass in this case. The last few lines of the previous set of commands, starting from the "compose" statement, are a sanity check on the symbolic solution in that they investigate what happens for the case in which $\theta = 0$. Here, the results of the level-ground solution are as follows:

$$a = \frac{g\mu_s d_1}{(d_1 + d_2) - \mu_s h}$$

Sample Problem 6.4

The uniform pendulum in the accompanying diagram is released from a horizontal position. Determine the motion of the pendulum. The motion is retarded by friction at the pin, which always opposes the motion.

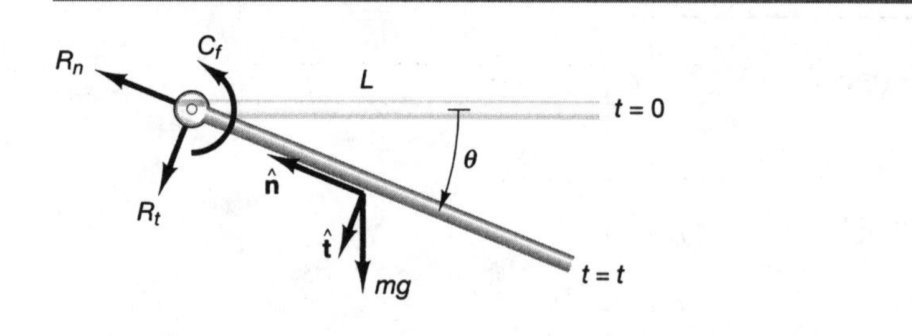

The solution to this problem requires numerical solution following the Euler method. The "ode" functions cannot be used on this problem, as the automated step size causes the matrix size limit of the student edition of MATLAB to be reached. The m-file containing the equation of motion and Euler integration is as follows:

```
x(1)=0;v(1)=0;t(1)=0;
dt=.01;
L=1;g=9.81;Mf=0.8;
for n=1:2000;
x(n+1)=x(n)+v(n)*dt;
v(n+1)=v(n)+((3/2)*(g/L)*cos(x(n))-Mf*sign(v(n)))*dt;
t(n+1)=t(n)+dt;
end
xd=x*180/pi;
plot(t,xd),title('theta in deg vs time in sec')
```

The following commands are entered to clear the values of x, v, and t and to run the file:

```
EDU>t=0;x=0,v=0;
EDU>n6pt4
```

This set of commands yields the following plot:

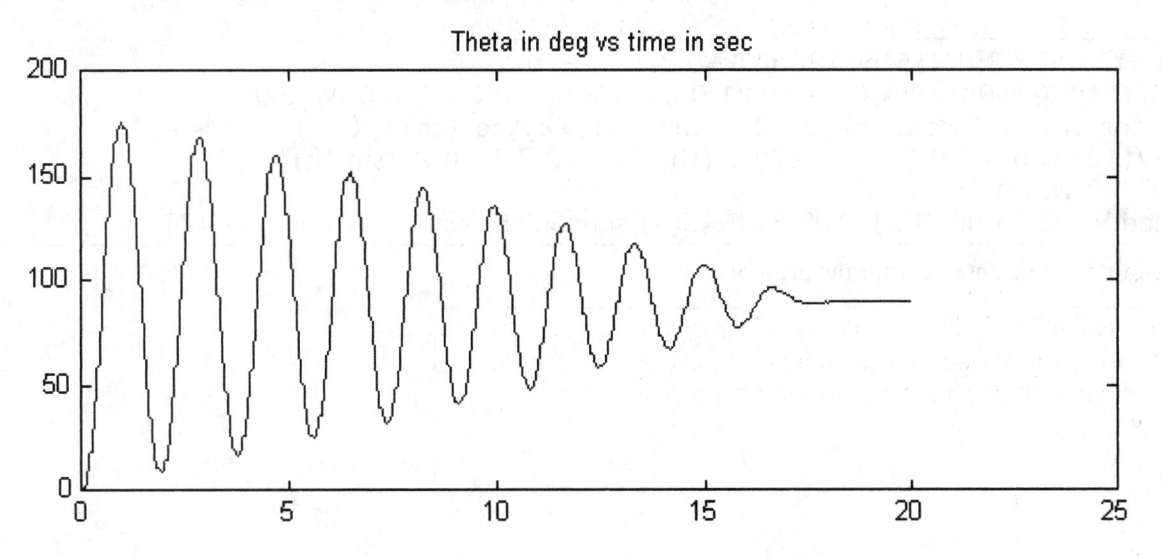

This simulation shows the oscillation dying out after 17 seconds because of the friction at the pivot, with a period of about 2 seconds.

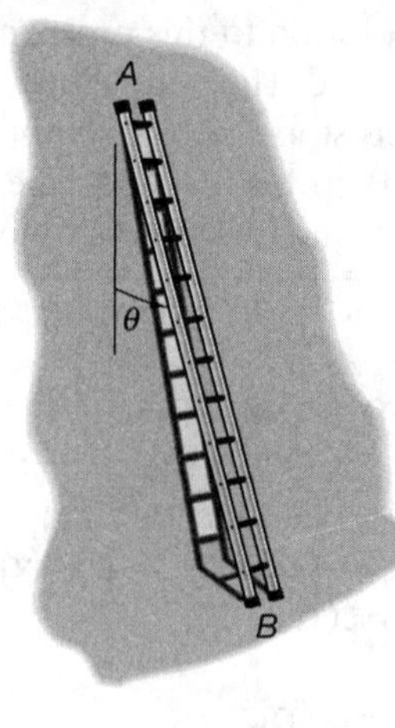

Sample Problem 6.5

Consider a uniform ladder (as shown in the accompanying diagram) of weight W that slips when it is leaned against a building:

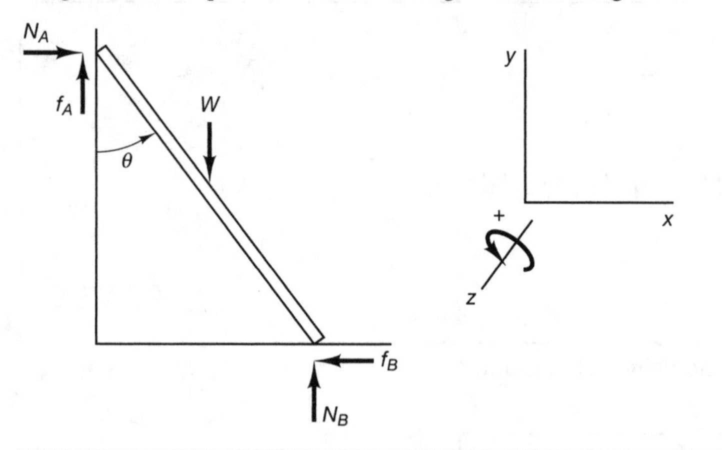

The solution to this problem is a straightforward use of matrix inversion and is set up in an m-file so that the various parameters can be easily changed:

```
W=50;L=12;g=32.2;th=35*pi/180;muk=0.2;
C=[-muk 0 1 0 0 0 0;0 -muk 0 1 0 0 0;1 0 0 -1 -W/g 0 0;0 1 1 0 0 -W/g 0;
-(1/2)*cos(th) (1/2)*sin(th) -(1/2)*sin(th) -(1/2)*cos(th) 0 0
-W*L^2/(12*g);0 0 0 0 1 0 -(1/2)*cos(th);0 0 0 0 0 1 (1/2)*sin(th)];
Load=[0;0;0;W;0;0;0];
S=C\Load;NA=S(1),NB=S(2),fA=S(3),fB=S(4),acmx=S(5),acmy=S(6),alpha=S(7)
```

The name of this file is entered after the prompt to yield:

```
NA =
   9.7638
NB =
   47.9216
fA =
   1.9528
fB =
   9.5843
acmx =
   0.1156
acmy =
   -0.0809
alpha =
   0.2822
```

Sample Problem 6.6

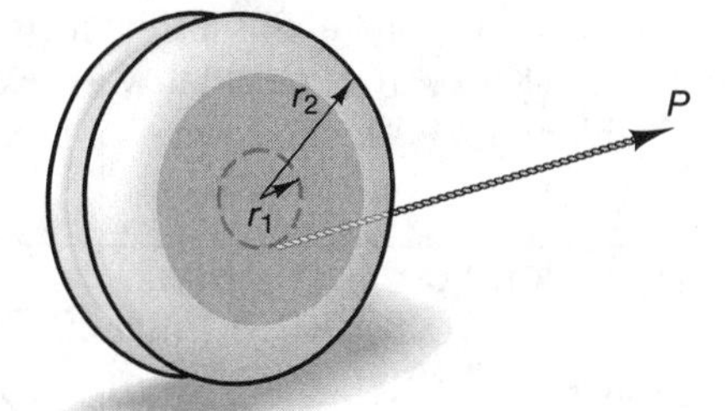

A yo-yo is rewound by pulling it across the floor by applying a constant force **P**, as illustrated in the accompanying diagram. If the inner radius of the yo-yo is r_1, the outer radius is r_2, and the radius gyration is k, determine the minimum coefficient of friction for the yo-yo to roll up the string without slipping.

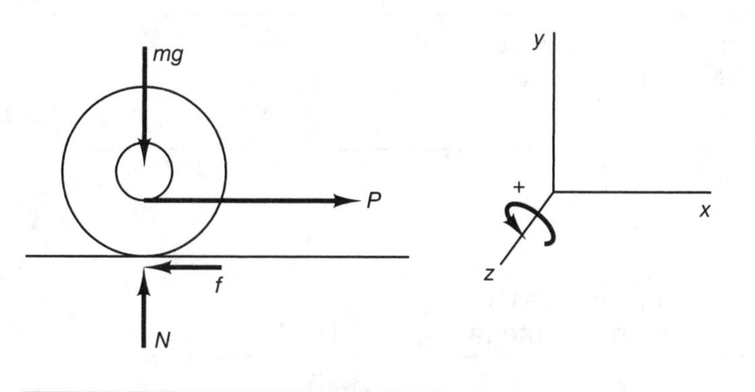

The calculation in this sample problem may be obtained by use of the following symbolic commands:

```
EDU>syms r1 r2 m k P;
EDU>A=[1 -m*k^2;r2 m*k^2];L=[P;P*r1];
EDU>S=inv(A)*L;f=S(1),alpha=S(2)
f
1/(1+r2)*P+1/(1+r2)*P*r1
alpha =
-r2/m/k^2/(1+r2)*P+1/m/k^2/(1+r2)*P*r1
```

For the yo-yo to roll without sliding requires that $f \geq \mu_S mg$. From the value of f that was computed symbolically, this requirement becomes

$$\mu_s \geq \frac{P(k^2 - r_1 r_2)}{mg(k^2 + r_2^2)}.$$

Sample Problem 6.8

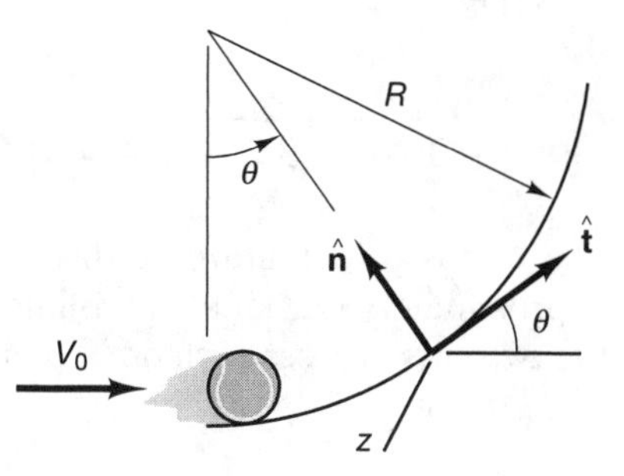

In Sample Problem 6.7, a ball was rolled up an inclined plane. If, instead, the ball is rolled up a curved surface (see the accompanying diagram), the differential equations will be nonlinear. In this case, normal and tangential coordinates are used to formulate the problem.

The solution to this problem is a straightforward application of the Runge–Kutta routine that was explained in the solution to Sample Problem 1.14. The following m-file, called sixpt8.m contains the equations to be solved:

```
function xdot=sixpt8(t,x);
R=100;r=0.5;g=9.8;,muk=0.2;V0=3;
xdot=[x(2);-(g/R)*(sin(x(1))+muk*cos(x(1)))-muk*x(2)^2;
     -(5*muk/(2*r))*(g*cos(x(1))+R*x(2)^2)];
```

Next, this m-file is called and solved using ode45 command, and the calculations and plots of the solutions are entered as follows in the command window:

```
EDU>tspan=[0 0.5];
EDU>x0=[0;3/100;0];
EDU>[t,x]=ode45('sixpt8',tspan,x0);
EDU>Rdth=100*x(:,2);rw=-0.5*x(:,3);
EDU>subplot(1,2,1),plot(t,x(:,1)),title('theta (rad) vs t (s)'),
subplot(1,2,2),plot(t,rw,'+',t,Rdth),title('detheta and -romega (+) vs t (s)')
```

These commands produce the following plots:

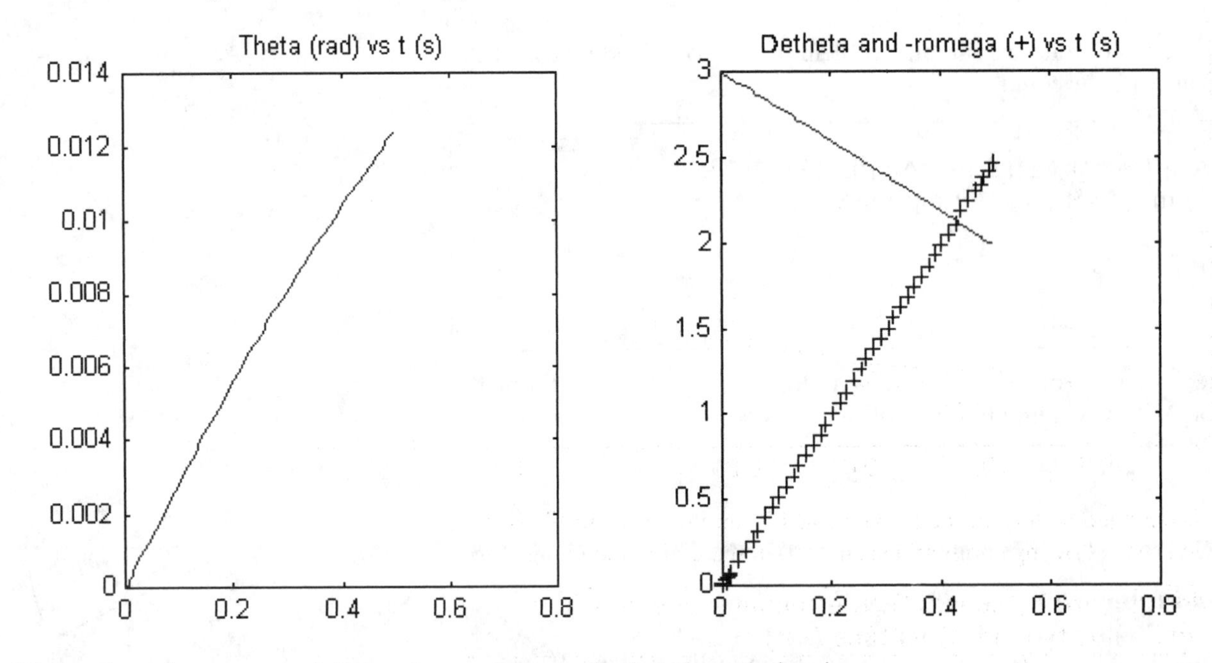

The radius of curvature of 100 m is very large such that the solution is approximately that of a ball rolling and sliding on a flat surface. Note that the two curves cross each other at a little past $t = 0.4$ sec.

Sample Problem 6.9

Consider the ladder sliding down a wall in Sample Problem 6.5. The acceleration of the ladder was determined for the instant that the ladder started to slip. The equations of motion were given as

```
function xdot=sixpt9(t,x);
L=12;g=32.2;m=0.2;
xdot=[x(2);3/((2-m^2)*L)*((1-m^2)*g*sin(x(1))
      -2*m*g*cos(x(1))+m*x(2)^2)];
```

Next, the following code is entered in the command window to solve and plot:

```
EDU>tspan=[0 1.6];
EDU>x0=[35*pi/180;0];
EDU>ode45('sixpt9',tspan,x0),title('angular position (rad)
      and velocity (rad/s) vs time (s)')
```

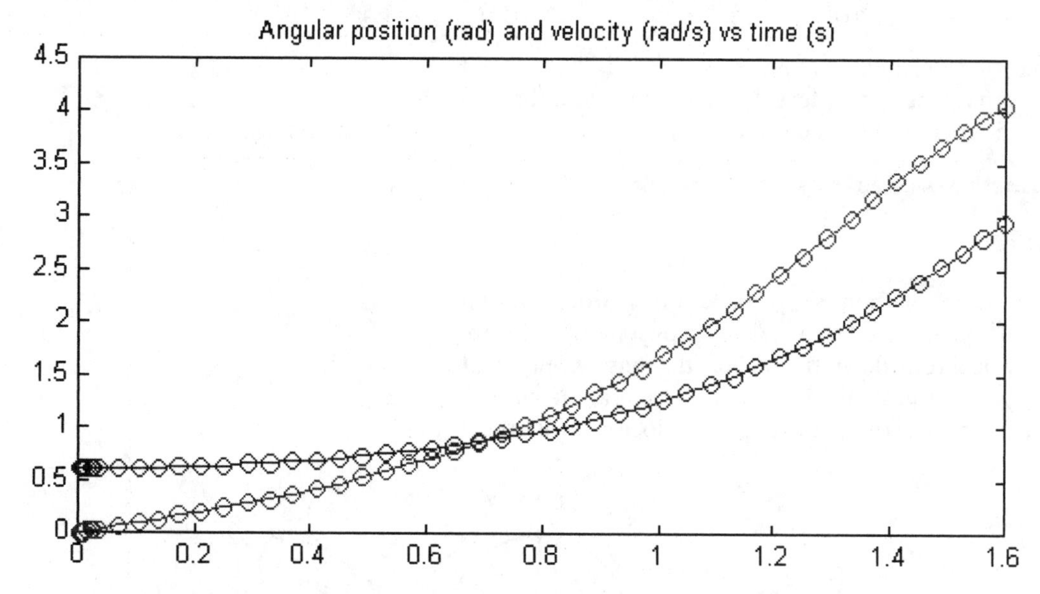

Now, μ_k is changed to zero in the previous m-file, and then the m-file is run again when the following command is entered in the command window:

```
EDU>ode45('sixpt9',tspan,x0),title('angular position (rad)
      and velocity (rad/s) vs time (s)')
```

This command results in the following plot:

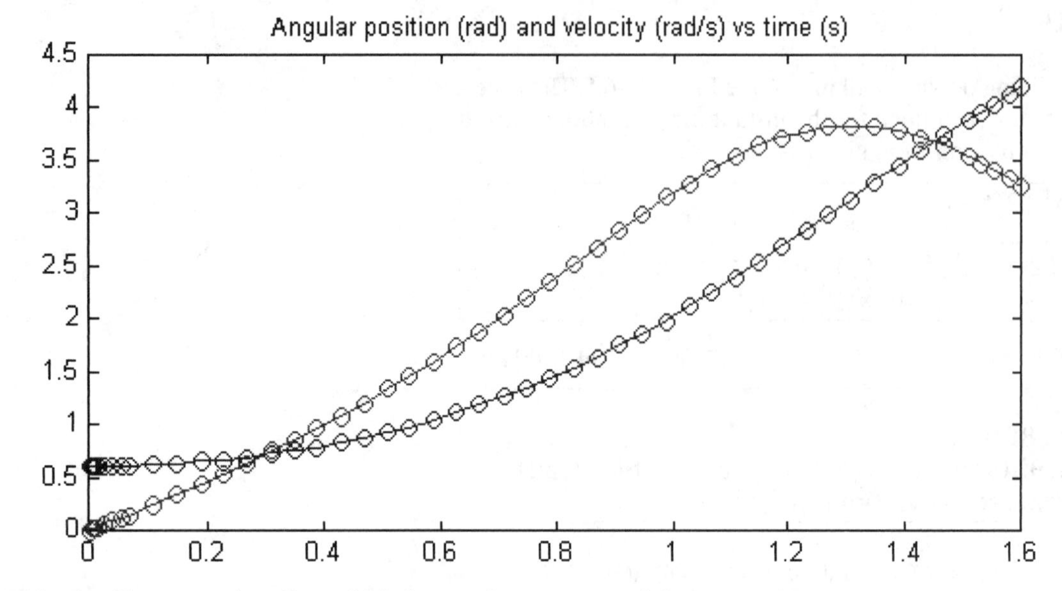

This plot illustrates the effect of friction on the response of the system. Note that the curves are over the same time interval and are subject to the same initial conditions; only the friction differs between them.

Sample Problem 6.11

A wheel rolls down an incline without slipping. A cord is wound around the outside of the wheel and attached to a mass as shown in the accompanying diagram. As the wheel rolls down the incline, the mass is pulled into the wheel. The system will be in a steady state of acceleration until the mass is completely drawn in. Determine the angular velocity of the wheel.

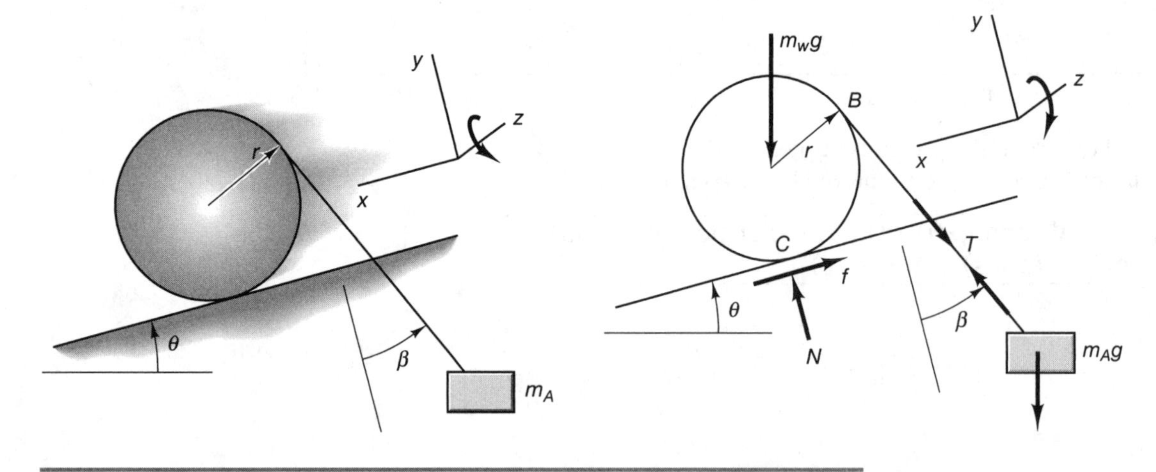

Solution of Nonlinear Algebraic Equations

MATLAB solves a system of nonlinear equations by using the symbolic manipulation command for small numbers of equations and by using an iterative technique (the Newton–Rapson method) for large numbers of equations. "fsolve", the command for numerically solving a system of nonlinear equations resides in the *Optimization Tool Box* and may not be available in the *Student Edition of* MATLAB. However, the fsolve function may be available if you are working from a university-site license, and it should be used if available. If it is not available to you at your institution, you should enter a short version of it that appears in the appendix under the name mnewt.m. Both of the commands mnewt and fsolve invoke Newton–Rapson methods for numerically solving systems of nonlinear algebraic equations. Such systems of equations have multiple solutions and require an initial guess of the solution nearest the solution of interest. Basically, the numerical procedure starts with an initial guess of the unknown variables and then continues to iterate until the solution converges with the inital guess. As might be expected, the success of the iterative technique is dependent upon the initial guess. The equations to be solved must all be placed in a form such that zero appears on one side of the equality.

If you have the student edition of MATLAB, you need to use the mnewt program provided in the appendix and explained on pages 26–27 of the *MATLAB Manual for Statics*. To use mnewt requires three different m-files. The first m-file contains mnewt.m, which you should copy from the appendix. (If you worked through the *MATLAB Manual for Statics*, you may have already saved this file.) The second m-file contains the constants, the names of the variables in the order that they appear in the vector "p" of initial guesses and the equations to be solved, which are placed in the vector "q".

```
function q=sixpt11(p)
g=9.81;mw=40;mA=10;th=30*pi/180;r=0.4;izz=0.5*mw*r^2;
T=p(1);f=p(2);N=p(3);b=p(4);acm=p(5);ax=p(6);ay=p(7);a=p(8);
q=zeros(8,1);
q(1)=mw*g*sin(th)-f-T*sin(b)-mw*acm;
q(2)=N-mw*g*cos(th)-T*cos(b);
q(3)=T*r-f*r-izz*a;
q(4)=T*sin(b)+mA*g*sin(th)-mA*ax;
q(5)=T*cos(b)-mA*g*cos(th)-mA*ay;
q(6)=acm+r*a;
q(7)=ax-acm+r*a*sin(b);
q(8)=ay+r*a*cos(b);
```

Next, create an m-file to call the Newton–Rapson method (again it is called through either fsolve or mnewt) and an m-file that contains the equations to be solved. The following m-file calls, solves, and displays the solution:

```
x0=[90;150;400;0.5;2;2.5;2;-5];% vector of initial guess
x=mnewt('sixpt11',10,x0,0.001,0.001);
fprintf('T = %8.3f\n',x(1));
fprintf('f = %8.3f\n',x(2));
fprintf('N = %8.3f\n',x(3));
fprintf('Beta = %8.3f\n',x(4));
fprintf('acm = %8.3f\n',x(5));
fprintf('aAx = %8.3f\n',x(6));
fprintf('aAy = %8.3f\n',x(7));
fprintf('alpha = %8.3f\n',x(8));
```

Saving this file as "sol6pt11.m" and then typing "sol6pt11" after the prompt yields the following solution:

```
T = 109.960
f = 150.472
N = 443.970
Beta = -0.327
acm = 2.026
aAx = 1.375
aAy = 1.918
alpha = -5.064
```

The fsolve and mnewt commands require that several pieces of information be provided when they are used. The first piece of information is the name of the m-file that contains the equations to be solved (in this case, the name is sixpt11), which must be set in single quotes and followed by a comma. Next, mnewt requires that the maximum number of trials allowed for the iteration be set. (In this example, the maximum is set at 10.) Such a number is not required by the more sophisticated fsolve command. Another required piece of information is the vector containing the initial guess (xo), which is to be placed between commas, and this value must be followed by the tolerance for the x computation and for the y computation (both are set to .001 in mnewt and to the default state, which is denoted by "[]" in "fsolve"). Finally, these values must be followed by a list of the constant parameters to be passed to the equations, each value separated from the others by commas.

The m-file used to call the mnewt command should be followed by eight fprintf commands. These commands are special output commands that take as arguments the listing of the name of a variable followed by an equals sign and then the format under which you want the computed value of the variable to appear. In this case, the designation %8.3 indicates that the value should be printed in a field that is eight characters long to a precision of three decimals. The f that follows indicates a fixed-point notation, and the "\n" indicates that the value should be printed on a new line. After the single quote and a comma, the name of the variable is given. So, for instance, if units, such as degrees, are required, they can be included between the f notation and the \n notation. More about this write-formated data command can be obtained by typing help fprintf at the prompt.

7

Power, Work, Energy, Impulse, and Momentum of a Rigid Body

Chapter 7 of the *Dynamics* examines the formulation of plane-dynamics problems using work–energy or impulse–momentum principles that are based on the first integrals of the motion. Thus, in general, the methods used in Chapter 7 do not require the solution of differential equations. However, MATLAB may be used to plot many of the solutions of the problems in Chapter 7 in order to visualize the motion and gain greater insight into the problems of dynamics.

Sample Problem 7.9 is set up as a general accident-reconstruction program, and the values of the initial velocities, the vehicle orientation, the normal vector to the damage area, and the coefficient of restitution are given. When commercial accident-reconstruction codes are used, engineers must vary all of the parameters until they are confident that they have the best fit for the available data. In this example, we assume that the preimpact data are known or estimated, and we need to calculate the postimpact data. The analysis in the Sample Problem in the Dynamics text is then used to determine the post-impact data, thereby establishing the initial velocities of the vehicles involved in the accident. In many circumstances, if one vehicle was exceeding the speed limit, the legal responsibility for the accident lies with the speeding driver.

Sample Problem 7.9

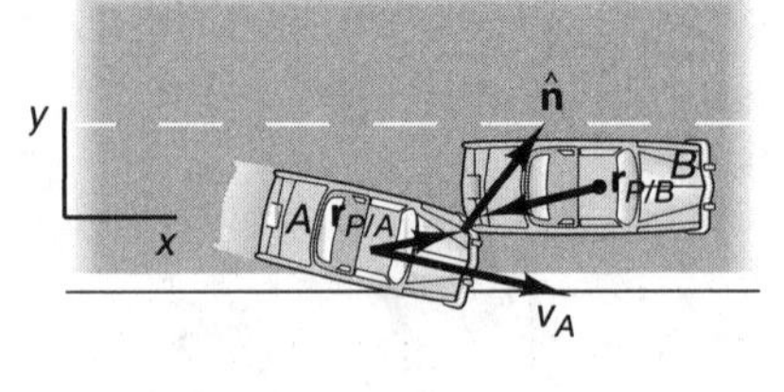

In the accompanying diagram, a Pontiac, vehicle B, is rear-ended by a Toyota, vehicle A, while waiting to make a left turn. Determine the postimpact dynamics if, while reconstructing the accident, you base your analysis on the following data: Witnesses report that the Pontiac was stopped waiting to make a left turn and the speed of the Toyota was 50 mph at the time of impact—that is, the Toyota did not brake before impact. This could be verified by the absence of skid marks. The direction of the velocity vector of A is assumed to be $-10°$ off the horizontal. The direction of the normal vector $\hat{\mathbf{n}}$ is assumed to be 5° from the horizontal. This direction can be obtained by examining the damage to the two vehicles. The coefficient of restitution is obtained by examining the damage to the vehicles and from the position of the vehicles after the accident is over. The values found for the various parameters are

$$m_A = 82.3 \text{ lb-s}^2/\text{ft} \qquad m_B = 87.6 \text{ lb-s}^2/\text{ft}$$

$$I_A = 1{,}686 \text{ lb-ft-s}^2 \qquad I_B = 1{,}795 \text{ lb-ft-s}^2$$

$$\mathbf{r}_{P/A} = 6\hat{\mathbf{i}} + 2\hat{\mathbf{j}} \text{ ft} \qquad \mathbf{r}_{P/B} = -7\hat{\mathbf{i}} - 2.2\hat{\mathbf{j}} \text{ ft}$$

$$\mathbf{v}_A = 73.3(\cos 10°\hat{\mathbf{i}} - \sin 10°\hat{\mathbf{j}}) \text{ ft/s} \qquad \mathbf{v}_B = 0$$

$$\omega_A = 0 \qquad \omega_B = 0$$

$$\hat{\mathbf{n}} = \cos 5°\hat{\mathbf{i}} + \sin 5°\hat{\mathbf{j}}$$

$$e = 0.8$$

In an actual reconstruction, the engineer should check for sensitivity of the solution to any variation in the direction of the unit vector $\hat{\mathbf{n}}$, the velocities, and the coefficient of restitution. The four equations for the postimpact normal component of the velocity and the angular velocity are solved using matrix notation.

This problem is solved by using an m-file so that the initial conditions can easily be changed to solve new problems with different initial data. The data is entered, and then various algebraic and vector calculations are made. These calculations are followed by the matrix solution for the post impact velocities and angular velocities in normal and tangential coordinates. These results are then combined to produce the values of the velocities of A and B after the collision. (Note that post impact quantities are denoted by "p" in the variable name.)

```
% first enter all of the initial conditions
thA=-10*pi/180;betn=5*pi/180;mA=82.3;mB=87.6;
IA=1686;IB=1795;wA=0;wB=0;
rPA=[6;2;0];rPB=[-7;-2.2;0];
% next calculate the various required values
vA=73.3*[cos(thA);sin(thA);0];
vB=[0;0;0];n=[cos(betn);sin(betn);0];k=[0;0;1];
t=cross(k,n);e=0.8;vAt=dot(vA,t);vAn=dot(vA,n);
vBt=dot(vB,t);vBn=dot(vB,n);
XA=dot(k,cross(rPA,n));XB=dot(k,cross(rPB,n));
% Now use these values to compute the postimpact values (denoted p)
A=[mA mB 0 0;XA*mA 0 -IA 0;0 XB*mB 0 -IB;-1 1 -XA XB];
F=[mA*vAn+mB*vBn;XA*mA*vAn-IA*wA;XB*mB*vBn-IB*wB;e*(vAn+XA*wA-vBn-XB*wB)];
S=inv(A)*F;vApn=S(1);vBpn=S(2);
wAp=S(3),wBp=S(4),vAp=vApn*n+vAt*t,vBp=vBpn*n+vBt*t
```

Typing the name of the m-file (sp7pt7) after the prompt yields the following results:

```
wAp =
    4.2330
wBp =
    4.2792
vAp =
   13.3976
  -17.8718
         0
vBp =
   55.2319
    4.8322
         0
```

Now, the postimpact analysis of vehicle A is continued. As the "ode" commands would exceed the size limit of the student version of MATLAB, an m-file is prepared with the physical constants, a simple Euler integration (see the discussion just prior to Sample Problem 1.6 in this supplement), and the equations to be integrated. These equations are very long and are thus split up into velocities and forces. Note that the dot operations are used because element-by-element operations are required.

```
m=82.3;I=1686;mu=0.7;g=32.2;L1=sqrt(2.5^2+3.2^2);
L3=sqrt(2.5^2+4.8^2);beta=atan(2.5/3.5);alpha=atan(2.5/4.8);
W1=0.3;W3=0.2;t=0;x1=0;x2=0;x3=0;x4=0;x5=0;x6=0;v1x=0;v2x=0;v3x=0;v4x=0;v1y=0;
v2y=0;v3y=0;v4y=0;
F1x=0;F1y=0;F2x=0;F2y=0;F3x=0;F3y=0;F4x=0;F4y=0;
dt=0.01;t(1)=0;
x1(1)=-pi/18;x2(1)=4.2;x3(1)=0;x4(1)=13.4;x5(1)=0;x6(1)=-17.9;
for n=1:150;
% Express the velocity of each tire as a function of the linear
% velocity of the center of mass, the angular velocity of the
% car, and the position vector from the center of mass of the tire.
v1x(n)=x4(n)-x2(n)*L1.*sin(x1(n)+beta);
v1y(n)=x6(n)+x2(n)*L1.*cos(x1(n)+beta);
v2x(n)=x4(n)+x2(n)*L1.*sin(x1(n)-beta);
v2y(n)=x6(n)+x2(n)*L1.*cos(x1(n)-beta);
v3x(n)=x4(n)-x2(n)*L3.*sin(alpha-x1(n));
v3y(n)=x6(n)-x2(n)*L3.*cos(alpha-x1(n));
v4x(n)=x4(n)+x2(n)*L3.*sin(alpha+x1(n));
v4y(n)=x6(n)-x2(n)*L3.*cos(alpha+x1(n));
% Now use these velocities to specify the
% force vectors acting on the tires.
F1x(n)=-g*mu*W1*v1x(n)./sqrt(v1x(n).^2+v1y(n).^2);
F1y(n)=-g*mu*W1*v1y(n)./sqrt(v1x(n).^2+v1y(n).^2);
F2x(n)=-g*mu*W1*v2x(n)./sqrt(v2x(n).^2+v2y(n).^2);
F2y(n)=-g*mu*W1*v2y(n)./sqrt(v2x(n).^2+v2y(n).^2);
F3x(n)=-g*mu*W3*v3x(n)./sqrt(v3x(n).^2+v3y(n).^2);
F3y(n)=-g*mu*W3*v3y(n)./sqrt(v3x(n).^2+v3y(n).^2);
F4x(n)=-g*mu*W3*v4x(n)./sqrt(v4x(n).^2+v4y(n).^2);
F4y(n)=-g*mu*W3*v4y(n)./sqrt(v4x(n).^2+v4y(n).^2);
% Now write the differential equation in terms of the previous velocities and forces.
x1(n+1)=x1(n)+x2(n)*dt;
x2(n+1)=x2(n)+(-g*mu*(m/I)*W1*(L1^2*x2(n)+L1*cos(x1(n)+beta).*x6(n)-
L1*sin(x1(n)+beta).*x4(n))./sqrt(v1x(n).^2+v1y(n).^2)-
g*mu*(m/I)*W1*(L1^2*x2(n)+L1*cos(x1(n)-beta).*x6(n)+L1*sin(x1(n)-
beta).*x4(n))./sqrt(v2x(n).^2+v2y(n).^2)-g*mu*(m/I)*W3*(L3^2*x2(n)-L3*cos(alpha-
x1(n)).*x6(n)-L3*sin(alpha-x1(n)).*x4(n))./sqrt(v3x(n).^2+v3y(n).^2)-
g*mu*(m/I)*W3*(L3^2*x2(n)-L3*cos(alpha+x1(n)).*x6(n)+L3*sin(alpha+x1(n)).*x
4(n))./sqrt(v4x(n).^2+v4y(n).^2))*dt;
x3(n+1)=x3(n)+x4(n)*dt;
x4(n+1)=x4(n)+(F1x(n)+F2x(n)+F3x(n)+F4x(n))*dt;
x5(n+1)=x5(n)+x6(n)*dt;
x6(n+1)=x6(n)+(F1y(n)+F2y(n)+F3y(n)+F4y(n))*dt;
t(n+1)=t(n)+dt;
end
```

Next, the name of the m-file is entered after the prompt in order to execute it, and then the following commands are typed in order to visualize the solution:

```
EDU>subplot(2,2,1),plot(t,x2)
EDU>subplot(2,2,1),plot(t,x2),title('Angular velocity vs time')
EDU>subplot(2,2,2),plot(t,x4,'+',t,x6),title('c of m velocities vs time')
EDU>subplot(2,2,3),plot(t,x1),title('Angular posiition vs time')
EDU>subplot(2,2,4),plot(x3,x5),title('C.G. displacement of car')
```

This code yields the following plots:

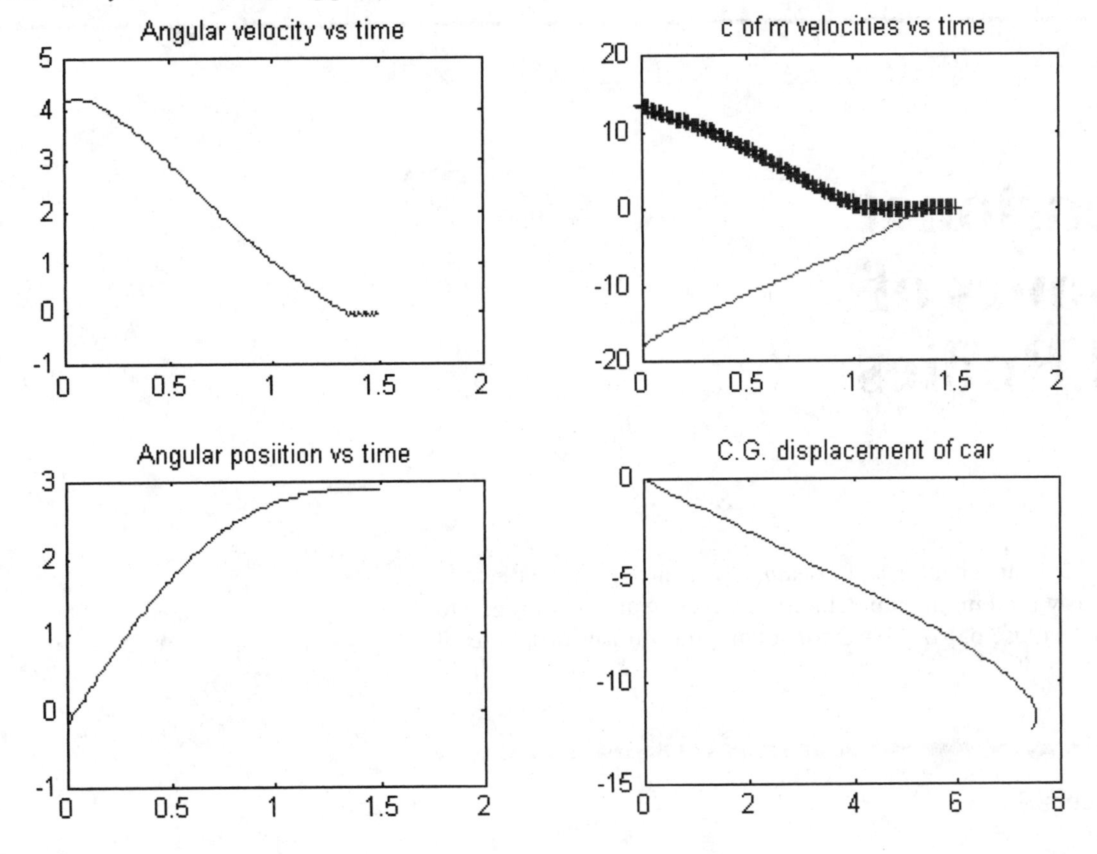

8

Three-Dimensional Dynamics of Rigid Bodies

The sample problems in Chapter 8 of *Dynamics* are all solvable with code that was previously used in this supplement. Again, you are encouraged to try to find your own use of MATLAB for solving the various problems of dynamics.

Sample Problem 8.3

Consider the transformation matrix given in Problem 8.8:

$$[R] = \begin{bmatrix} 0.353 & 0.918 & -0.177 \\ -0.353 & 0.306 & 0.884 \\ 0.866 & -0.25 & 0.433 \end{bmatrix}$$

Determine the Eulerian angles using Eq. (8.34).

The solution to this sample problem requires the use of two m-files and follows the steps of the solution to Sample Problem 6.11, which also involves a system of nonlinear algebraic equations. In this problem, however, the equations to be solved are overdetermined as there are five equations in only three unknowns. If you have not solved Sample Problem 6.11, you may need to refer back to its solution in this supplement, or to the appendix at the end of this supplement in order to create the m-file called mnewt.m. The first m-file contains the names of the variables in the order that they appear in the vector "p" of initial guesses, and the equations to be solved, which are placed in the vector "q".

```
function q=eightpt3(p);
th=p(1);phi=p(2);xi=p(3);
q=zeros(5,1);
q(1)=cos(th)-0.433;
q(2)=sin(th)*sin(phi)-0.866;
q(3)=-sin(th)*cos(phi)+0.25;
q(4)=sin(xi)*sin(th)+0.177;
q(5)=cos(xi)*sin(th)-0.884;
```

Next, an m-file is created to call mnewt (or fsolve, if you have access to it) and the previous file and to solve the algebraic equations:

```
x0=[pi/3;pi/3;-.1];
x=mnewt('eightpt3',10,x0,0.0001,0.0001);
fprintf('Theta=%8.3f\n',x(1));
fprintf('phi=%8.3f\n',x(2));
fprintf('xi=%8.3f\n',x(3));
```

Typing the name of this file after the prompt yields the following results:

```
Theta=   1.123
phi=   1.290
xi=  -0.198
```

Next, a solution near a different zero is sought by going back to the last m-file and changing the initial guess. For this solution, the m-file has been modified so that the answers appear in degrees rather than in radians, as they did in the previous solution. The modified m-file becomes:

```
x0=[-pi/3;20*pi/18;pi];
x=mnewt('eightpt3',10,x0,0.0001,0.0001);
r(1)=x(1)*180/pi;r(2)=x(2)*180/pi;r(3)=x(3)*180/pi;
fprintf('Theta=%8.3f\n',r(1));
fprintf('phi=%8.3f\n',r(2));
fprintf('xi=%8.3f\n',r(3));
```

Typing the name of this m-file after the prompt (or using the "save" and "execute" commands on a Mac OS) yields the following results:

```
Theta= -64.344
phi= 253.897
xi= 168.678
```

Both of the solutions agree with those found in the *Dynamics* text, but are obtained with less hand calculation.

Solution to the Heavy-Top Problem

The heavy axisymmetric top is discussed and presented on page 552 of the *Dynamic* text.

The solution to this requires numerical solution, which has been used very frequently throughout the previous solutions in this supplement, starting with the solution to Sample Problem 1.14. The next m-file prepares the equation of motion for this problem to be solved in first-order form using an Euler method so that the constant generalized momentum terms (constants of the motion or first integrals) can be imposed as the equation of motion is solved. The variables are defined as follows: $x1 = \theta$, $x2 = d\theta/dt$, $x3 = \phi$, $x4 = d\phi/dt$, $x5 = \psi$, and $x6 = d\psi/dt$.

```
% Euler integration for the heavy-top problem
m=0.5;L=0.02;Ix=0.001;Iz=0.0002;g=9.81;
a=(Iz/Ix)*(3*cos(pi/18)+100);
b=3*(sin(pi/18)^2)+a*cos(pi/18);
N=2000;dt=0.001;        % set the number of iterations and step size of the time
x1=zeros(1,N);x2=x1;x3=x1;x4=x1;x5=x1;x6=x1;t=zeros(1,N);    % clear and size all vectors
x1(1)=pi/18;x2(1)=0;x3(1)=0;x4(1)=3;x5(1)=0;x6(1)=100;    % set the initial conditions
c=Ix*a/Iz;
for n=1:N;
x1(n+1)=x1(n)+x2(n)*dt;
x2(n+1)=x2(n)+((1/Ix)*m*g*L*sin(x1(n))+x4(n).^2.*sin(x1(n)).*cos(x1(n))-
a*x4(n).*sin(x1(n)))*dt;
x3(n+1)=x3(n)+x4(n)*dt;
x4(n+1)=((b-a*cos(x1(n)))./(sin(x1(n))).^2);
x5(n+1)=x5(n)+x6(n)*dt;
x6(n+1)=(c-(b*cos(x1(n))-a*cos(x1(n)).^2)./(sin(x1(n))).^2);
t(n+1)=t(n)+dt;
end
x1d=x1*180/pi;x3d=x3*180/pi;
subplot(1,2,1),plot(t,x1d),title('nutation angle')
subplot(1,2,2),plot(t,x3d),title('precession angle')
```

Typing the name of this m-file in the command window yields the following plots:

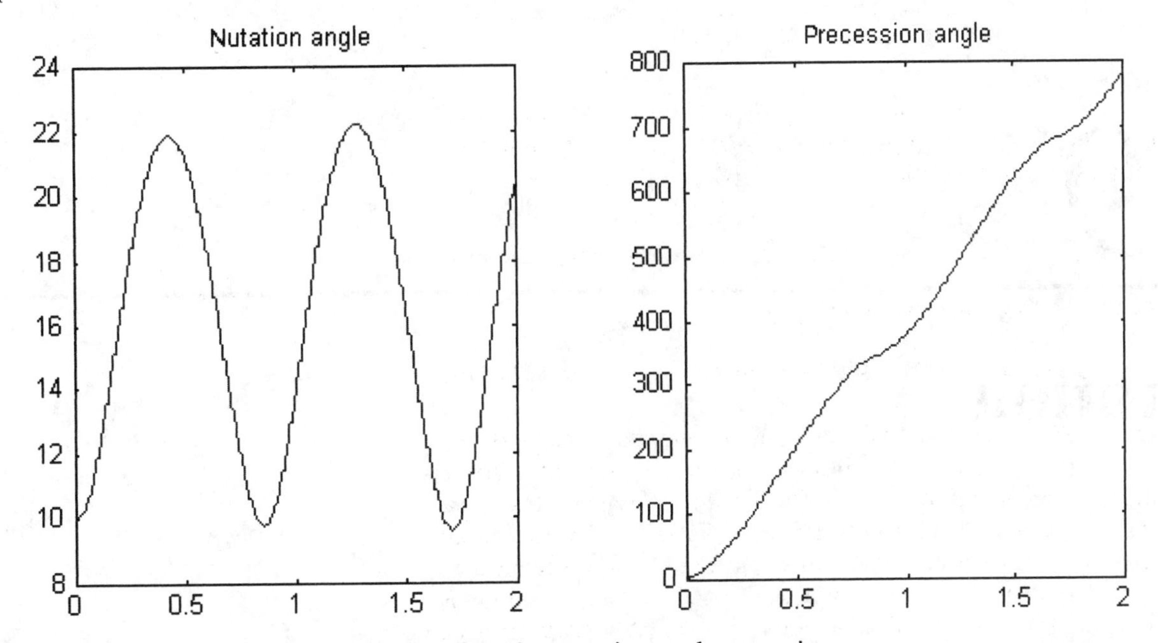

Next, it is a simple matter to examine the nutation and precession angles for various choices of physical examples by simply changing these values in the m-file. For example setting new values of $I_{xx} = 0.6 \times 10^{-6}$, $I_{zz} = 0.2 \times 10^{-3}$, $m = 1$, and $L = 0.03$ and changing the initial value of $d\psi/dt$ to 120 rad/s yields the following plots (it must not be forgotten that all vectors should be set to zero before the program is run again):

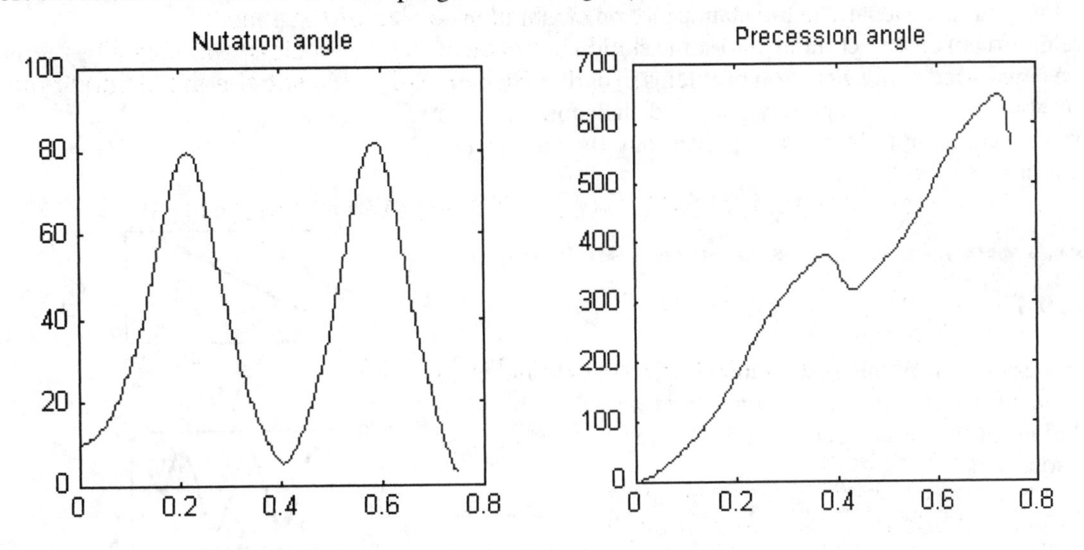

9

Vibration

Chapter 9 of *Dynamics* considers a particular dynamic motion of vibration—that is, a repetitive motion of an object relative to some stationary frame of reference. We have considered many vibration problems in earlier chapters of this supplement, but have not completely analyzed the motion. Since vibration problems are common in industrial applications, they are examined in detail in this chapter.

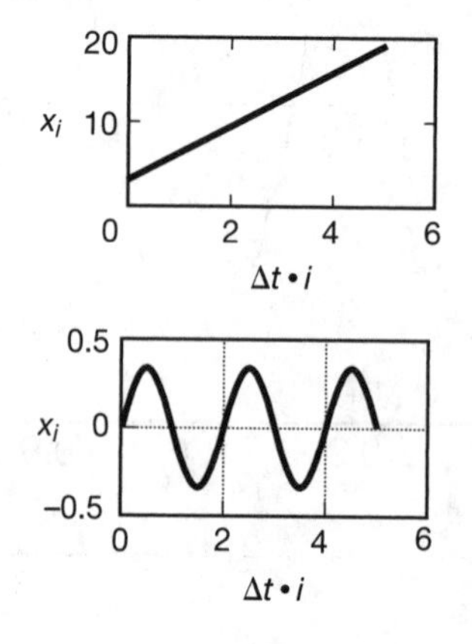

Sample Problem 9.6

Compute and plot the response of the pendulum of the Sample Problem 9.5 if $g/L = 10$ and the initial conditions are $\theta = \pi$ rad and $\dot{\theta} = 1$ rad/s. Repeat the calculation for the initial conditions $\theta = 1$ rad and $\dot{\theta} = 1$ rad/s. Compare the two solutions.

The differential equation of motion for the pendulum is

$$\ddot{\theta} = -\frac{g}{l}\sin\theta,$$

with the following sets of initial conditions: a) $\dot{\theta} = 1$ rad/s and $\theta = \pi$ rad or b) $\dot{\theta} = 1$ rad/s and $\theta = 1$ rad.

The m-file for case a) is

```
% Pendulum equation via the Euler method
x(1)=pi;v(1)=1;t(1)=0;
dt=0.001;
for n=1:10000;
x(n+1)=x(n)+v(n)*dt;
v(n+1)=v(n)-10*sin(x(n))*dt;
t(n+1)=t(n)+dt;
end
plot(t,x)
```

This m-file produces the following plot of the angle versus the time for the first 10 seconds:

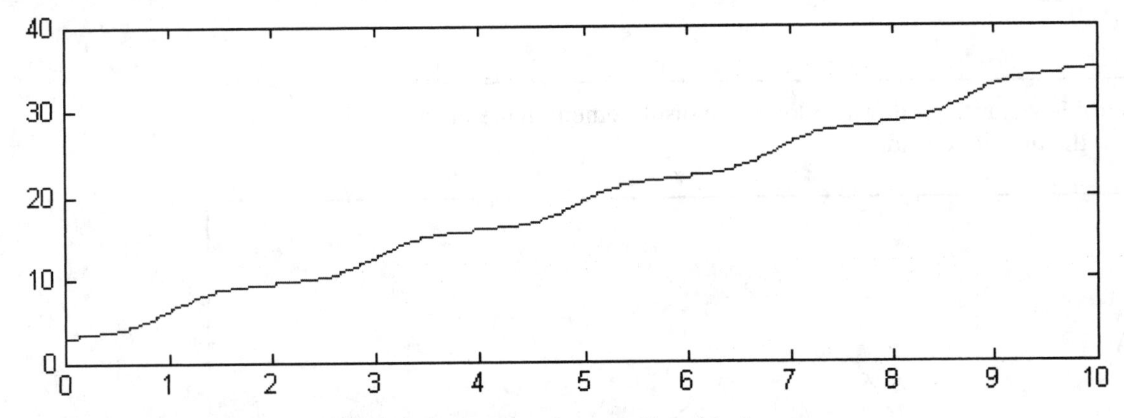

Now, case b) is considered by running the previous m-file with the new initial conditions. The following plot of angular position versus time results after the m-file is run:

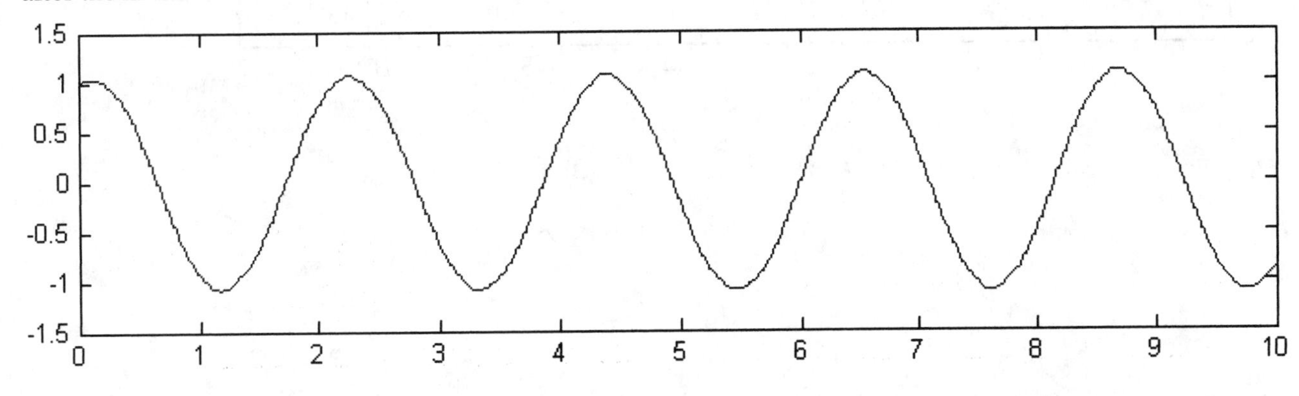

Sample Problem 9.9

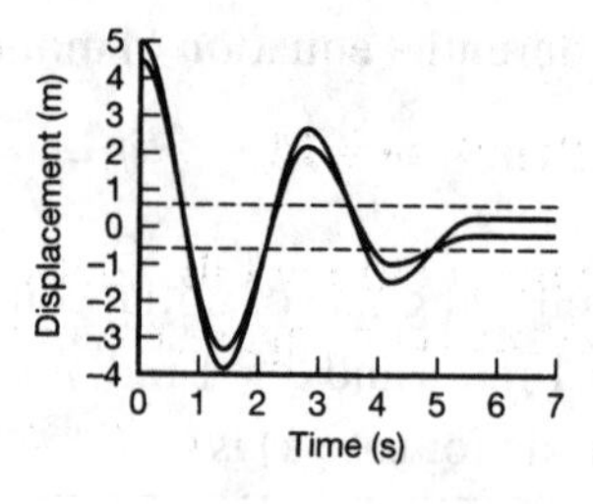

Compute and plot the response of the system of Figure 9.9 with a coefficient of friction of $\mu = 0.3$, mass of $m = 100$ kg, and stiffness of $k = 500$ N/m, for the two different initial conditions (a) $v_0 = 0$ and $x_0 = 4.5$ m and (b) $v_0 = 0$ and $x_0 = 5.0$ m.

The differential equation is numerically solved using the Euler method. The following m-file defines the equation to be solved, the initial conditions, and the Euler solution:

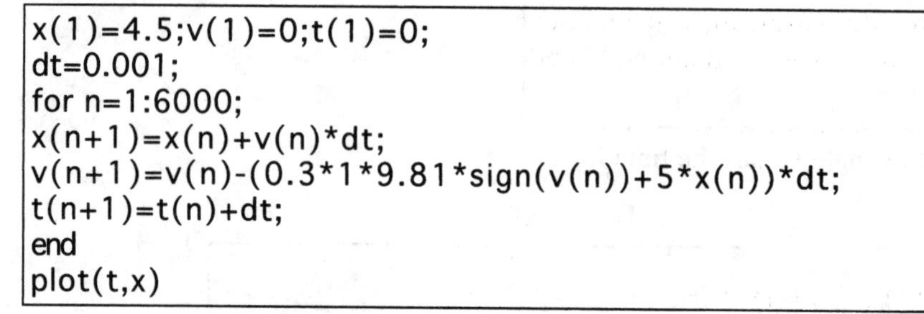

```
x(1)=4.5;v(1)=0;t(1)=0;
dt=0.001;
for n=1:6000;
x(n+1)=x(n)+v(n)*dt;
v(n+1)=v(n)-(0.3*1*9.81*sign(v(n))+5*x(n))*dt;
t(n+1)=t(n)+dt;
end
plot(t,x)
```

Running this m-file yields the following plot of the displacement in meters as a function of the time in seconds:

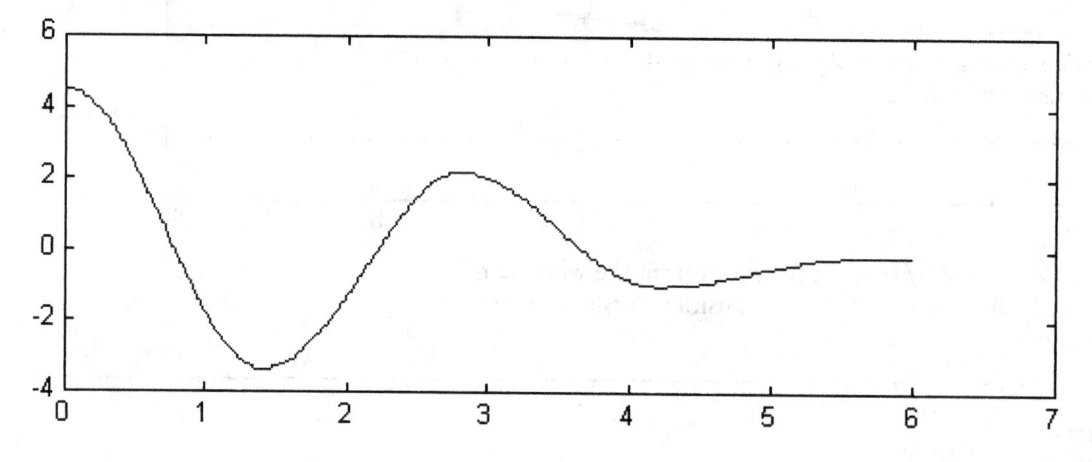

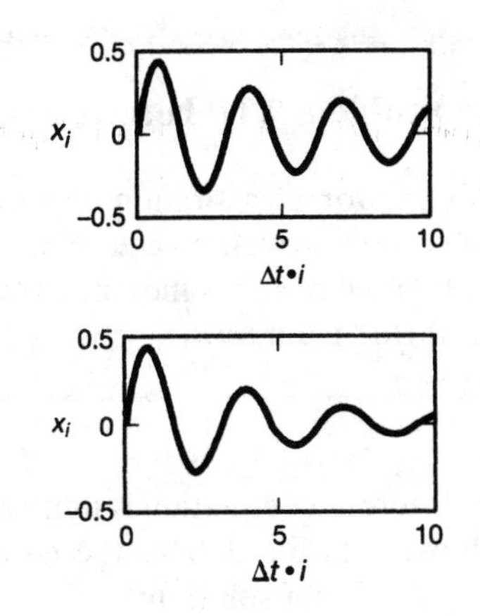

Sample Problem 9.10

Compute the solution of the following system, which models damping due to air viscosity acting against a spring–mass system, and plot the result:

$$m\ddot{x} + c\dot{x}|\dot{x}| + kx = 0$$

Here, $m = 50$ kg, $k = 200$ N/m, $c = 25$ kg/s, $x_0 = 0$, and $v_0 = 1$ m/s. Compare this result with that obtained from a system with linear viscous damping with the same damping coefficient.

The differential equation is numerically solved using the Euler method. The following m-file defines the equation to be solved, the initial conditions, and the Euler solution:

```
x(1)=0;v(1)=1;t(1)=0;
dt=0.001;
for n=1:10000;
x(n+1)=x(n)+v(n)*dt;
v(n+1)=v(n)-(0.5*v(n)*abs(v(n))+4*x(n))*dt;
t(n+1)=t(n)+dt;
end
plot(t,x)
```

Running this m-file produces the following plot of the displacement in meters versus the time in seconds:

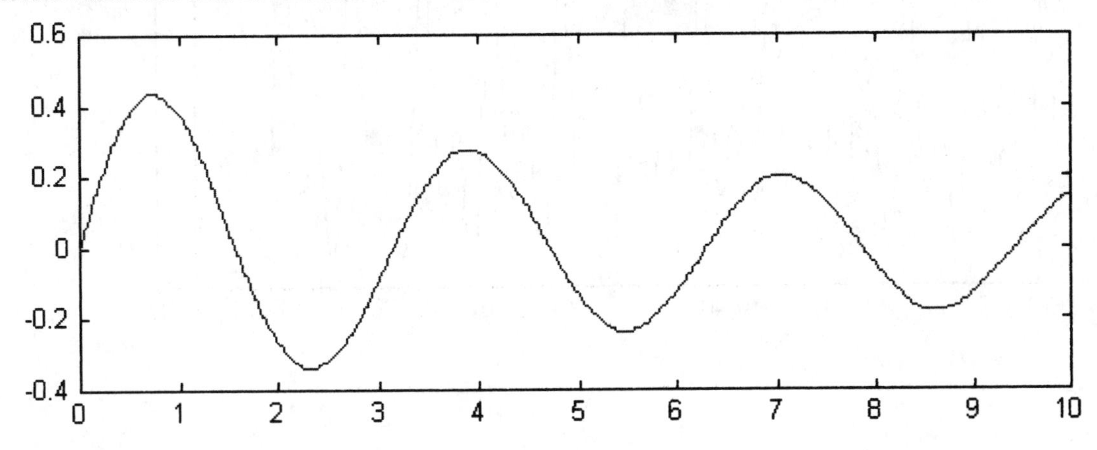

Sample Problem 9.11: Forced Response and Resonance

Consider the forced vibration of a mass m connected to a spring of stiffness 2,000 N/m, driven by a 20-N harmonic force at 10 Hz. The maximum amplitude of vibration is measured to be 0.1 m, and the motion is assumed to have started from rest ($x_0 = v_0 = 0$). Calculate the mass of the system.

The differential equation is numerically solved using the Euler method. The following m-file defines the equation to be solved, the initial conditions, and the Euler solution:

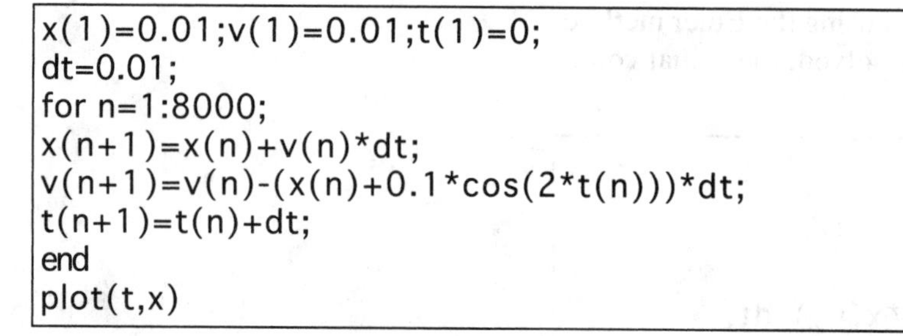

```
x(1)=0.01;v(1)=0.01;t(1)=0;
dt=0.01;
for n=1:8000;
x(n+1)=x(n)+v(n)*dt;
v(n+1)=v(n)-(x(n)+0.1*cos(2*t(n)))*dt;
t(n+1)=t(n)+dt;
end
plot(t,x)
```

Running this m-file produces the following plot of the displacement in meters versus the time in seconds:

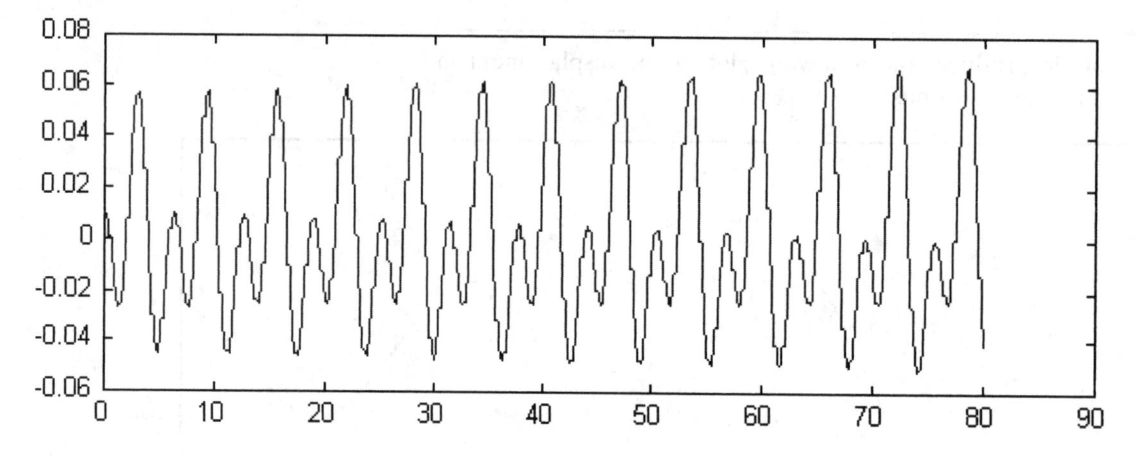

Appendix A

MATLAB is very useful for computing principal moments of inertia, particularly if they are formulated as an eigenvalue problem, as illustrated in the following command window:

```
EDU>I=[10 -5 3;-5 8 4;3 4 7];    % enter the matrix I
EDU>det(I)   % compute the determinant of the matrix I
ans =
 33
EDU> [V,D]=eig(I)   % compute the eigenvalues and eigenvectors of I
V =
    0.7580   -0.4151   -0.5032
   -0.6506   -0.4264   -0.6283
   -0.0462   -0.8037    0.5933
D =
   14.1090        0         0
         0  10.6718         0
         0        0    0.2192
```

Note that the "[V,D]" command returns the eigenvalues as the diagonal elements of the matrix **D** and the eigenvectors as the columns of the matrix **V**. Next, computation of the roots to the cubic equation for the principal moments shall be considered by using the polynomial root function in MATLAB. This approach represents the polynomial by entering the coefficients of the polynomial in descending order.

```
EDU>P=[1 -25 156 -33];     % enter the coefficients
EDU>beta=roots(P)     % solves for the roots of the polynomial
beta =
   14.1090
   10.6718
    0.2192
```

Next, some additional commands are included to plot the polynomial, providing a visualization of the three roots. This task is done by first defining a range of values of x using the linspace command. Next, the polyval command is used to evaluate the previously defined polynomial (saved as the vector "P") and store it in the vector v. The last command plots v, which represents $f(\beta)$, versus "x", which represents β (in kg · m^2).

```
EDU>x=linspace(0,15);
EDU>v=polyval(P,x);
EDU>plot(x,v),title('cubic equation for principal moment')
```

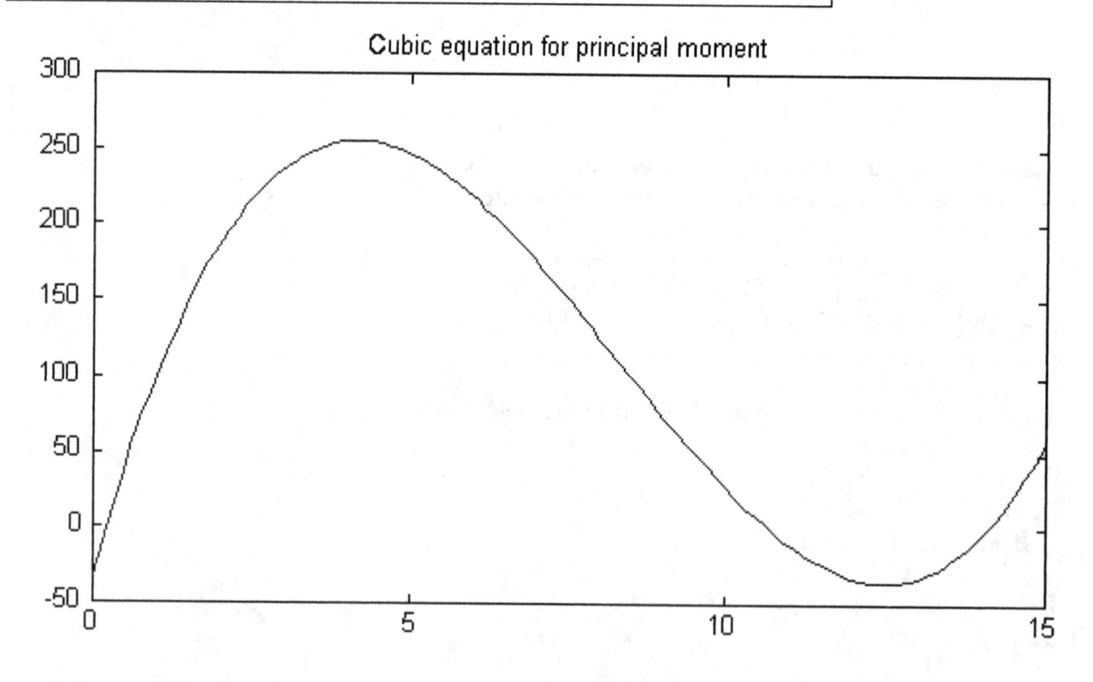

Appendix B: The "mnewt" Program

This appendix contains the code for solving a systems of nonlinear equations using a Newton–Rapson method for those users who do not have access to the fsolve function. The fsolve command is found in the Optimization Toolbox as an optional attachment to the professional version of MATLAB. The code for "mnewt" is a short version of the code for "fsolve" and should be suitable for the problems encountered in statics and dynamics. The "mnewt" command is designed to be used in much the same way as the fsolve command is used.

```
function a = mnewt(FUN,ntrial,guess,tolx,tolf,varargin)
% MNEWT Solve multiple nonlinear equations using a Newton-Rapson method
% MNEWT(FUN, ntrial guess, tolx, tolf, parameters)
% FUN - name of the function to be tested; function returns
% results in a column vector
% ntrial - maximum number of trials
% guess - initial guess
% tolx - tolerance in the iterative step size
% tolf - tolerance in function results
% parameters - parameters that will be passed directly to FUN
% after the guess
% by Cole Brooking, The MathWorks, Inc. 7/3/97
for k = 1:ntrial
 beta = feval(FUN,guess,varargin{:});
 for i = 1:length(guess) % Compute partials
 guess2 = guess; delta = abs(guess(i)*1e-5);
 guess2(i) = guess2(i)-delta;
 alpha(:,i) = (feval(FUN,guess2,varargin{:}) - beta)./delta;
 end
 errf = sum(abs(beta)); if errf < tolf; break; end;
 beta = alpha\beta; errx = sum(abs(beta)); guess = guess + beta;
 if errx < tolx; break; end;
end
a = guess;
```

Some MATLAB Commands Commonly Used in Dynamics

`sin(x)`	compute the sine of the number x
`exp(x)`	compute the exponential of the number x
`cos(x)`	compute the cosine of the number x
`sqrt(x)`	compute the square root of the number x
`linspace(1,10,9)`	creates a vector of 9 elements evenly spaced, starting with 1 and ending with 10
`length(x)`	returns the length of the vector x
`size(A)`	returns the length of the vector x
`dot(x,y)`	computes the dot product of the two vectors x and y
`cross(x,y)`	computes the cross product of the two vectors x and y
`norm(x)`	computes the Euclidian norm of the vector x
`A'`	computes the transpose of the array A
`det(A)`	computes the determinant of the square matrix A
`inv(A)`	compute the inverse of the matrix A
`eig(A)`	compute the eigenvalues and eigenvectors of the matrix A
`syms('x')`	declares that x is a symbolic variable
`diff(f)`	differentiates f symbolically or if f is a vector of numbers diff takes the difference between its elements
`solve(f-10)`	symbolically solves the equation f = 10, f a symbolic expression
`int(f,s)`	symbolically integrates the function f with respect to s
`ezplot(f,[a,b])`	plots the symbolic function f between the numbers a and b
`subs(f,x,s)`	substitute x by s in the symbolic expression f
`compose(f,g)`	computes f(g(x)), f and g symbolic functions of x
`ode45('name',t,x0)`	returns the numerical solution of the first order differential equation stored in the `name.m` over the time interval saved in t, starting with the initial conditions stored in x0, and plots it.
`plot(t,x)`	plots x as a function of t where x and t are the same size vectors
`plot3(a,b,t)`	constructs a 3d plot of the vectors a and b as functions of t
`quad8('name',a,b)`	computes the integral of the function stored in the file `name.m` between the numbers a and b

INDEX